TAKEOFFS AND LANDINGS

TAKEOFFS AND LANDINGS

With gratitude to those who, chronologically starting in 1938, encouraged me most in an often seemingly impossible journalistic venture; namely, that if pilots understood their airplanes better, as well as the causes of most accidents, there'd be few accidents and general aviation would develop much more rapidly. With a common desire to lend a hand these good people made all the difference:

W. T. PIPER • RALPH MCA. INGERSOLL • ARTHUR GODFREY • REED M. CHAMBERS • GERARD B. LAMBERT, JR. • CHARLIE MILLER • CHARLES FROESCH • A. C. SIMMONDS, JR. • ED AND LIB BRADEN • DEWITT WALLACE • FRANKLIN T. KURT • GEORGE E. HADDAWAY • WALLACE E. PRATT • WM. D. STROHMEIER • JEROME LEDERER • R. C. OERTEL • ROBERT N. BUCK • WOLFGANG LANGEWIESCHE • DWANE L. WALLACE • JACK GATY • LEDDY GREEVER • NORMAN WARREN • WM. P. LEAR • BEVERLY E. HOWARD • HENRY B. DUPONT • CARL FRIEDLANDER • CHARLES A. HINSCH • WM. A. MARA • FRED E. WEICK • HOWARD PIPER • W. T. PIPER, JR. • THOMAS F. PIPER • J. WILLARD MILLER • JOE DIBLIN • DONALD C. FLOWER • ELIZABETH GORDON NORCROSS • MOSS PATTERSON • BILLY PARKER • ALBERT L. HARTING • DAVID DOWS • JOHN A. STEELE • O. A. BEECH • ROSCOE WILSON, M.D. • DWIGHT P. JOYCE • PHIL MCKNIGHT • FRANK E. HEDRICK • D. H. HOLLOWELL • R. JAMES YARNELL • ROBERT L. CHATLEY• KEN BRUGH • OSCAR MEYER, JR. • JUDGE B. REY SCHAUER • HARRY L. KIRKPATRICK • O. C. KOPPEN • BETTY RYAN WOLFE • W. B. LENKARD • JOE GEUTING • GEORGE R. GALIPEAU • H. D. VICKERS, M.D. • GEORGE HUGHES • JIM RIDDLE • RUDY GARFIELD • WM. ENYART • LES WALLACK• JO KOTULA • JOHN K. THORP • J. H. LAPHAM • LARRY HIRSCHINGER • HOWARD S. STERNE • ALFRED M. BERTOLET • S. J. MANETTE • LLOYD O. YOST • KENNETH LITTLE • A. CLAYTON TSCHANTZ • LARRY ZYGMUNT • DUANE STRANAHAN • DAVID D. BLANTON • RALPH HARMON • E. B. JEPPESEN • KARL FRUDENFELT, M.D. • RICHARD D. MORGAN • BILL MAULDIN • MYRON H. BUSWELL • MOULTON B. TAYLOR • JACK SCHOLEFIELD • LORD AND LADY CASEY • BRIG. GEN. CARL I. HUTTON • T. H. DAVIS • GUY HAM • CARL ALLY • JIM GREENWOOD • RICHARD E. YOUNG • BILL ROBINSON • EDWARD J. KING JR. • EDWARD D. MUHLFELD• WM. E. KELLEY • DAVID R. ELLIS • JAMES F. NIELDS

CONTENTS

List of Photographs

FOREWORD

You, too, must have had this experience—perhaps in respect to flying, or to your profession or to a sport. You meet someone who is an acknowledged master. You feel—if I could only make him talk! If he were a computer, I'd push the button and make him "dump"—print out all that's in him. What a gold mine that would be! But he is a person. You can't just say "tell me all." You have to ask questions. And I don't even know what to ask him. That's half the problem: What do I need to know?

Well, for flying, here the thing is done. A smart woman, Eleanor Friede, a pilot herself, who owns a Grumman American Traveler, asked Leighton Collins one simple question: "Would you tell us all you know about takeoffs and landings?" The result is now before you, and it is a gold mine.

Who *is* Leighton Collins? Most pilots know anyway, but here are the main facts. He first soloed on open-cockpit biplanes. In '29, he bought his first airplane, a Cardinal, a side-by-side cabin high-wing, fast for its time. He flew it coast to coast and all over in between, on business as an insurance underwriter. In 1933 of all times—the economy had ground to a halt—he became, of all things, an airplane salesman. He would take an airplane from the factory and not come back till he had sold it. That sometimes would take weeks. He would drop in at some airport, most any airport, and offer demonstration rides to local pilots, most any pilot. Since nobody had any money, there was no point in discriminating between prospects. And sometimes the local renter pilots might talk the local operator into buying a "Bathtub" Aeronca, so they could fly it at perhaps three dollars an hour. . . .

It took, on the average, 100 demonstration flights to sell one airplane. Demonstrating was not the same as instructing. You put your prospect into the left-hand seat right from the start, let him go ahead, and kept your mouth shut. Collins rode with literally thousands of pilots of the most varied experience, skill, and personality. That's where he learned about what is now called "the man-machine system": the pilot and the airplane, and how they get along with one another. What he found was not good.

Pilots lacked skill and understanding. They had no idea where the risks lay. They jittered about engine failure when they should have worried about their own reaction to a stall or spin. Many had white knuckles. Some were so scared they trembled. Others tried zany things. One pilot, at 500 feet in the hot and fast little Monocoupe, opened the throttle wide and accelerated to top speed. That time Collins didn't keep his mouth shut. "What are you going to do?" he asked. "Loop," the prospect said. "Not with me," said Collins. . . . And airplanes were not right. Many had tricky ailerons. On the turn to final, in slow flight under G-load, the left wing would go down, the pilot would try to pick it up with the ailerons and this alone could trigger a spin.

Collins would ask himself: "Will this man kill himself?" and quite often decided "probably." Sooner or later he would stall at low altitude, probably in a turn; the nose would go down and he would see the ground come at him. At that moment would he push or would he pull? Too often it was clear that he would pull.

Coming back to some airport a year later, Collins would sometimes find that some former prospect was now dead. Almost always the fatal accident had developed out of turning flight at a low altitude.

Leighton Collins decided to do something about all this. In February '38, he brought out the first issue of a monthly bulletin, *Air Facts*. It was at first nothing but an account of recent fatal accidents, in full grim detail. The idea was shocking. Accidents were not supposed to be discussed in print. They were best swept under the rug. "Airmindedness" must not be discouraged. Private persons were not supposed to nose around. The government did investigate but did not publish its findings in form usable for pilots. *Air Facts* pioneered the accident report which is now so common.

As a publishing idea, it seemed suicidal—selling people a detailed account of their own prospective demise! But people were smarter than people thought. Almost right away 3000 subscriptions came in; this at a time when there were only 15,000 licensed pilots in the United States.

Air Facts was pocket-size. As a business it was vest-pocket size. Collins was chief pilot, chief investigator, writer, editor, publisher, production manager, circulation manager, and accountant; later, when ads came in, also chief space salesman. He typed his own pieces. Even in later years the staff never numbered more than four. His first office, in the old General Motors Building in New York, was so small that visitors just stood in the hallway and talked through the open door. But they came. Captains of industry and finance came—the few who at that time owned or flew airplanes. Airline captains dropped in to exchange information. Hollywood stars walked in. Old Man Piper came. An early issue had made it clear that the Cub not only could spin and kill its occupants but that it quite often did. A distributor had sent Mr. Piper a copy, saying: "If you can't get *that* stopped, we'll never sell another Cub." Piper, being the man he was, gave a year's subscription to *Air Facts* to every purchaser of a Cub for a while—until the war training boom started. The list of subscribers was always small but very select—a veritable Who's Who of flying in America, salted with a few kings, princes, foreign heads of state. The subscribers' fleet was all the best that was flying.

Air Facts soon branched out into other features—new-airplane critiques, airplane-in-use stories, and pieces on flying technique. That, too, was pioneering. At that time, what a pilot really did with his controls was not considered something you could much discuss in print. If you can't *feel* it, forget it!

But *Air Facts*'s most appealing feature was Leighton's monthly account of his own latest trip. Somehow he always found a week each month to fly far and wide, probing the weather, the minds of pilots, and (as he once put it), the small airplane's vinelike way of growing into our lives. Somehow *Air Facts* always yielded him a good airplane; at times a second one which contributors got to fly. How he got the tiny business to yield so much I've never understood. Perhaps it was his early training at the Harvard Business School. He certainly understood his business more clearly than many big corporations seem to. Perhaps his secret was in keeping it small: no use publishing for people who don't want the information anyway. *Air Facts* was really a research institute that supported itself and paid taxes too. It was an extraordinary feat for 34 years.

In the early period great effort went into one question—how do you get through under the weather? At that time there was no practical way to put real instruments on small airplanes beyond a venturi-driven turn and bank. There were no suction pumps, often no generator. Horizons or gyros were then as radar and deicing are now—you did not put that sort of thing on that sort of airplane. The economics forbade. Effective transmitters, too, were too expensive. Only the airlines flew instruments, and the military sometimes. For the rest of us, flying was "contact" willy-nilly. And the "contact" was close: 500 and one was considered quite okay. The question was—could you get through still lower?

Could you find special bad-weather routes? Could you perhaps mark such routes? Were there regional rules-of-thumb, local weather wisdom that could help you? Once he ran totally out of weather, in snow showers. All he saw was a red barn, and a field near it that was good only for an easy crack-up. He flew back and forth across the barn for an hour or so, and the weather passed.

When war surplus instruments came in, things changed— slowly. Collins was one of the first to push small airplanes IFR, at first on a venturi-driven directional gyro. He never advocated flying IFR. He just did it, and his detailed accounts encouraged many others. And as always, he told the facts, good or bad. Many pilots still considered it Russian roulette to go IFR in a single-engined airplane. Well, sure enough— one time he was on solid instruments, carrying considerable ice. His fuel vent (invisible from the cockpit) iced up, and his one-and-only engine quit. He made a power-off instrument approach at Dayton, breaking out over the Middle Marker. The manufacturer found a new ice-free location for the vent, and that was that. Had the airplane had two engines, they would both have quit.

Then came helicopters. He became a sophisticated helicopter pilot and, at one time, the oldest active helicopter pilot. When *Air Facts* moved to Princeton, he had his own heliport behind his house. And then came the exploration of the light twin, of which you'll find a distillate in this book.

As an analyst of flying, Collins differs from all others in his sharp focus on the realities. To go by most books and manuals (including mine), the airplane is an idealized machine, all airplanes more-or-less alike. It is flown in an idealized air— generally smooth—by an idealized pilot who is cool, rational, well-informed. In reality every model of airplane is different, often significantly so, and no airplane is ever perfect. The perfect airplane cannot exist. Collins points out that airplanes have to be built for the real market—what cannot be sold cannot be built. So we have airplanes that, in landing, won't let you look where you want to look; airplanes that present you with stick-forces, perhaps due to downsprings, that interfere with control-feel; airplanes which on a crosswind takeoff get light on their wheels before they are ready to fly and skitter sidewise over the ground. And so on. Because in the real market an airplane must carry as much as possible, we have airplanes that can be misloaded with the CG too far aft or too far forward and in that condition they can turn around and bite. And we have the light twins which, with one engine out, demand more pilot skill and self-control than many pilots are prepared to furnish.

As for the pilot, he is not the rational operator whom the manuals imagine. He wants to prove he is good. He wants to show off. Sometimes he's simply in a hurry! In some situations he is beset by animal fear. He wants to get through. His judgment often misses the point. Decades ago Collins told me: "There is no weather problem, VFR. You can always quit before you run out of VFR. There's always an airport nearby. But pilots won't quit. The weather problem is really a utilization problem, and an ego problem."

As for the real air, Collins saw decades ago a problem that has only recently moved into official cognizance—wind shear, the structure of gusts, their effect on the airplane in climb and descent and in turns. "The theory of flight," he said, "is written for smooth air. But we are playing around in the surf."

In 1972 Leighton Collins sold his *Air Facts.* But not to retire! He next launched into another large flying program as a consultant for a manufacturer. He explored the many, often subtle design features that make pilots get along better with one airplane than with another. And then came Eleanor Friede with her innocent question—how, really, do you take an airplane up and get it down again in one piece? And so half a century of close observation and penetrating analysis have gone into this book. We are fortunate to have it. Read it slowly. There is a lot to get.

—*Wolfgang Langewiesche*
PRINCETON, N.J.
MAY 1981

INTRODUCTION

The title of this book may suggest that it deals with a limited even though prideful part of our flying. After all, how much of our flying time is spent in takeoffs and landings? Ten percent? Maybe less, hardly any more. But during this limited time, we produce roughly 70 percent of the reportable accidents each year. That's worth looking into.

As to the 30 percent of all accidents that occur out of the pattern and in the en-route phase, the bulk of these misadventures and catastrophes are weather related and lead into meteorological discussions that are beyond the purview of this book. It is well to note, however, that most of the weather and assorted en-route accidents are amenable to prevention by abstinence.

VFR pilots do not have to get caught on top of an overcast; they have to get there intentionally. Low- and high-time VFR, and occasionally IFR, pilots fly low in restricted-visibility conditions and hit whatever gets in front of them—mountainsides, trees, towers. Intended VFR flights start with the weather bad locally and known to be bad en route and at the destination; yet the pilots go and often, of all times, at night.

How can they be so bold, or optimistic, or misinformed? VFR pilots should never lose sight of the ground, but if they do and get on instruments, they are almost certain to get into a spin or spiral dive, the latter being the producer of structural overstress failures.

It is surprising how many pilots run out of gas flying cross-country, some even on an IFR flight plan, and also at night. This adds to the engine-failure and -malfunction list, which is already replete with easily avoidable carburetor-ice incidents.

Finally, no one has to engage in buzzing or low-altitude aerobatics, which have an 80-percent fatal rate if anything at all untoward happens.

The salient point about the en-route-accident picture is that such accidents reflect mainly gross errors in judgment. And are easily avoidable.

In contrast, there's little room for abstinence or avoidance in takeoffs and landings. If we're going to fly at all we've got to get it into the air somehow, and once we're airborne there's no way to avoid a landing.

For better perspective, in round figures the statistics on the fixed-wing, under-12,500-pounds-gross aircraft, exclusive of those involved in crop dusting and taxiing accidents, that will be involved in reportable accidents in an average year in general aviation run like this:

	Total Accidents	% Fatal or Serious Injuries
In Takeoff Roll or Aborts	250	3.6%
In Initial Climb	400	39.6%
In Traffic Pattern/Circling	80	55.0%
In Final Approach	420	30.9%
In Level-off/Touchdown	550	4.7%
In Landing Roll	550	.9%

So. In facing our takeoff-and-landing future, we need to recognize that this phase of our flying is the most demanding of the pilot and to think in terms of what sound basic concepts, piloting skills, and good procedures can provide us.

Contact?

1

BASIC CONCEPTS

It would be convenient to say that takeoffs and landings are affected by three things—aircraft flight characteristics, atmospheric conditions, and pilot flight characteristics—and go on from there. But it is not that simple. The variations from these three sources are so tightly intertwined in so many of our flights that a better starting point is a review of some of the more important basic concepts, and particularly how we perceive and try to control angle of attack.

It is very straightforward to think of elevator movement as causing a wing to pitch about its lateral axis, or of aileron displacement as causing it to roll about its longitudinal axis, or of rudder displacement as causing it to yaw about its vertical axis. But lateral, longitudinal, vertical—you can see that since we're accustomed to operating with at least one foot on the ground, our control movements tend to be ground referenced.

Stick or control wheel back and the nose rises above the horizon. Stick or wheel to the right and the right wing goes down toward the ground and the left one rises. Right rudder, say, and the nose swings or yaws to the right with reference to some point on the distant horizon. In many situations, for instance in a straight-ahead climb or descent, or in correcting a displacement of the airplane in level flight, these reactions serve us well enough.

But we also fly round the bend, and next are thinking about using the controls to make the airplane go where we want it to go. And we're off to the races. Sometimes we get into trouble less than halfway around the bend. Because, in rough air, we are called upon to exercise simultaneously pitch, roll, and yaw control. In our turns we need to think in terms of how our three controls enable us to rotate the airplane about its own pitch, roll, and yaw axes rather than about ground-referenced axes.

Meanwhile, in our turns and sometimes even in straight flight, we tend to forget that the first prerequisite to keeping an airplane flying is our ability to keep the angle at which the wing pushes through the air within very narrow limits. Too large an angle and the wing quits flying: the airplane becomes just so much weight with no means of support, and we get into gravity's 23 feet-per-second acceleration routine, which is vertically ground referenced.

The Wright brothers were not only the first, but the first to know all about angle of attack. With their pusher Flyer, they had a place out front where they could attach a string, which was free to align itself with the relative wind. They soon learned that in a proper climb or glide, the string would tilt, tail end up, aligning itself at a certain angle to the longitudinal axis of their machine. If they attempted to climb too steeply or glide too slowly, the string's angle would increase and they'd be in trouble because with the relative wind at that angle the wing would lose its ability to provide lift.

Later, when they departed from the straight and narrow and started exploring the turn phenomenon, they discovered that in a turn if they let (with elevator) the string's angle get higher than was proper in a climb or glide, they were in the same trouble as in straight flight, even though in the turn they might be flying faster than in a climb.

Today, these many years later, we have no string to tell us when we're flying too close to stalling angle of attack. So let us enter the labyrinth of how we do our best to keep our angle of attack within its small viable limits, using airspeed, attitude, power, control feel, and any other straw that flies by to give us a hint. And often agreement is not unanimous about how we even use or should use our three controls.

ELEVATOR VERSUS THROTTLE

As you know, in recent years there has been a lot of discussion about what controls airspeed and what controls altitude. By FAA ukase we are required to think, or at least to say on our writtens, that the elevator controls altitude and the throttle controls airspeed.

I think this is a horribly dangerous concept when the chips are down because it can cause pilots to pull back on the stick to go up, or, after an engine failure, to stay up. It is instinctive for people to think in terms of pulling back on the stick to climb or maintain altitude because it is so logical.

The best example I know of this universal impulse came a few years back. My friend, little Helen Langewiesche, had been given a spin demonstration before soloing a glider. One evening after dinner with Helen and her test-pilot/author father, Wolfgang, I asked her what she would do with the stick if ever she found herself nose-down and starting to autorotate. Her answer: "Why, dummy, I'd pull the stick back all the way because if you don't get the nose pointed up instead of down, you're going to fly into the ground." *Stick and Rudder,* over on the other side of the fireplace, became rigid. I paled. In the ensuing silence Helen added, "Oh, well, yes, first it is necessary to lower the nose to get some speed, and then you can get headed away from the ground." In short, she knew better, but her initial statement revealed a universal instinct that has to be trained out of people.

A proper concept of airspeed and altitude control is important in thinking about takeoffs and landings because these are low-altitude operations, and virtually all stall/spin/mush accidents begin close to pattern altitude. And those who, ground-shy, instinctively pull the stick back to gain or maintain altitude, go down.

THE STICK DOES BOTH

Yes, the stick controls altitude. When you're cruising along level, if you pull the stick back a bit, the altitude increases. No doubt about it. This is imprinted in the pilot's mind hundreds of times on even a relatively short flight. But in the process of increasing altitude, pulling back on the stick also decreases airspeed. On the other side of the coin, while flying along level in cruising flight, if you put some forward pressure on the stick, the altitude decreases. But note also that the airspeed increases. So, the stick controls both altitude *and* airspeed.

This is true for the simple reason that there is no such thing as a single-engine airplane. Every airplane has a second or additional engine: gravity. Along with its attitude and all-important angle-of-attack control functions, the stick is also the "throttle" for this additional engine.

Someday, with your airplane in landing configuration with full flaps, get the power set so that the airplane will fly at a constant altitude at normal approach speed. It will probably take about 50-percent power to accomplish this. Now gradually close the throttle while lowering the nose to the first mark below level on the artificial horizon and note the over-the-nose attitude. It will be about 10° nose-down. You'll find the airspeed in this power-off glide to be virtually the same as in level flight with 50-percent power. Your friendly gravity engine, which has never had an engine failure, can provide you with 50 percent of the power of your regular engine anytime you need it if you simply put the airplane in a 10° nose-down attitude. (Lift is always perpendicular to the relative wind and in a glide it has a thrust vector, provided by gravity.) And if you're ever in a bind for a rapid acceleration, be mindful that the nose doesn't have to go a lot lower than 10° for gravity to provide thrust equal to that of your regular engine. And this is done with the elevator control.

As well as providing a downhill thrust vector, gravity also provides an uphill drag factor or an increase in drag. Think in these terms for the moment to get a full grasp of the gravity effect. Go back to level flight, landing configuration, with 50-percent power set up. Now raise the nose almost to the first mark above level in the artificial horizon, which will be about your normal climb attitude with full power in takeoff configuration. You may find that even with full power, you can't quite maintain your level-flight speed. Gravity is pulling you back as much as the increase in power is pushing you forward, so there is no acceleration. Obviously gravity power, which is always there, can be a blessing or a curse. Gravity power: stick forward, more power; stick back, reverse thrust.

The dividing line is whether, with wings level, the nose is above the horizon or below the horizon.

THE JOB OF THE THROTTLE

What about the official concept, namely that the regular throttle controls airspeed? On takeoff, it is certainly convincing: as the throttle is advanced, the airplane starts moving along the runway and the airspeed starts moving right on up.

At this point the throttle is controlling airspeed. No doubt about it. But bear in mind that in the takeoff run, you're "flying" what is really only a ground-based vehicle. The rules of aerodynamics become applicable only at the moment of lift-off. An airplane's wing lives and breathes airspeed and flies only when it is moving faster than its applicable stall speed.

Actually, the best working postulate appears to be that the throttle controls altitude, not airspeed. In a climbout with full power, the thing that produces altitude is rate of climb. The highest possible rate of climb is available when the airspeed, controlled by the elevator, reaches the speed at which there is the least drag, or in other words, the aircraft's maximum rate of climb speed.

Now reduce power a little and lower the nose as necessary to hold the best climb speed. The rate of climb, or production of altitude, declines. Reduce power still more—the nose has to be lowered still more to hold the maximum rate of climb speed, and the rate of climb decreases further. So what the throttle produces is not airspeed but rate of climb, and it thus provides altitude control on the basis of whatever power is available in excess of what it takes to fly level at minimum airspeed. In a full-power climb at maximum rate of climb speed, there's no way to increase airspeed with throttle because there's no additional power available. As a normally aspirated airplane goes up, this power decreases and the nose has to be carried lower to hold the maximum rate of climb speed and consequently gain the maximum of altitude from the climb power available.

THE THROTTLE DOES BOTH

The most convincing demonstration—at least according to them—by the throttle-airspeed advocates needs close examination. In trimmed cruising flight, they go from cruise power to full throttle and maximum RPM and invite attention to the increase in airspeed. What they fail to mention is that as the speed increases, they have to apply increasing forward pressure on the stick, or add nose-down trim, so that the increased power will convert to airspeed rather than rate of climb. Otherwise, the airplane would simply not have increased its airspeed but would have climbed at the airspeed at which it was trimmed to fly. So, like the elevator and stick, the throttle also has a dual function: it controls both airspeed and altitude.

THE RULE WHEN YOU'RE PLAYING FOR KEEPS

While we may conduct experiments demonstrating to ourselves that elevator and throttle each control both airspeed and altitude, when the chips are down and altitude is scarce, a firm grasp of the basic concept that elevator controls airspeed and throttle controls altitude is critical. It's just hard to climb with the nose up and no power, but more than that I'm thinking of the poor fellow who has a partial power loss in a climbout, or, more rarely, a complete engine failure, with the throttle all the way in. His FAA-prescribed training and misinterpreted experience shouts to him, "The elevator controls altitude." And he stalls the airplane's wing. He ought to have kept his primary attention on maintaining flying speed.

Sometime, if occasion permits, at one of those government-sponsored how-to-fly seminars, ask the shouting, lectern-pounding evangelist what he pushes to increase his airspeed when flying a glider. The glider pilot has to have power to gain altitude, and he gets it from up-currents in the air while controlling his airspeed with the stick.

The stick-forward more-airspeed concept is also a livesaver in flying's most insidious trap: loss of control in turning flight at low altitude, as in a turn to crosswind after not quite leveling off from takeoff climb, or in an overtightened turn onto final to correct an overshoot of the runway center line. This loss of control occurs in a stall that, as will be developed farther on, may not always be entirely the fault of the pilot.

WHAT'S THE RUDDER FOR?

It may seem a bit late, in this day and age, to suggest that this question is of primary importance in getting our basic concepts straight. But understanding the rudder is vital, not only for en-route flying but in takeoffs and landings.

The rudder has been around for a long time in the marine world, where it is used to steer a boat and to turn it. Many people come to flying with that basic concept and really never get rid of it in the air. In an emergency seeming to require a quick reversal of course, they lead with and hold rudder in the turn, have to use opposite aileron to keep from overbanking, and stall and roll out the bottom of a skidded, crossed-control turn.

"What's the rudder for?" That exact question was once asked in one of Professor Otto Koppen's aeronautical engineering classes at MIT. His answer was a priceless précis: "The purpose of the rudder is to cover up the designer's mistakes."

What Professor Koppen meant was that the purpose of the rudder is to overcome undesired yaw or to produce desired yaw. In his reference to designers' mistakes, I'm sure he had primarily in mind the common practice of the times to use, for economic reasons, ailerons that, in rolling into a turn, caused the nose to swing away from the turn. There are still some in production today.

RUDDER COORDINATION

Adverse aileron yaw, from way back, provided a fetish for the instruction profession. Ability to counteract yaw when rolling into or out of turns became the hallmark of a smooth pilot. The way to smoothness: coordination of rudder and aileron movement.

In proportion to the designer's mistakes, coordination is important for maximum performance and better aiming in combat flying and in aerobatics and in safer everyday flying. To turn left, for instance, use left aileron and feed in just enough left rudder to keep the ball in the middle until the desired degree of bank is reached. Then release pressure on the controls. To come out of the left turn, use right aileron and right rudder as necessary to keep the ball in the middle until the wings are level. Or, in the absence of a ball, coordinate with the rudder just enough to avoid any feel of skidding or slipping while entering or coming out of turns, or in other words, while using ailerons to cause the airplane to roll.

The need for rudder coordination for smooth entry into and recovery from turning flight buttresses the concept that somehow the rudder has a primary function in causing an airplane to turn. And this is wrong.

OTHER ROLES OF THE RUDDER

Adverse aileron yaw characteristics of airplanes are not the whole story. We also have airplanes in which, through differential aileron travel, the up-aileron moves farther up than the down-aileron moves down. In consequence, the up-aileron produces just enough drag to offset the drag of the lift-producing down-aileron. The Cessna 152-2 and the Mooney ailerons work this way. Other airplanes—the Cessna 340 and even the single-place Pitts, for instance—use Friese-type ailerons, which operate in a slot and are hinged so that the leading edge of the up-aileron digs down into the airstream beneath the wing, creating enough extra drag to offset the drag of the down-aileron. In still others—the Bonanza, for instance—there is a connecting spring between the aileron and rudder cables, and displacement of the ailerons causes just enough automatic rudder movement to overcome the adverse yaw caused by the difference in drag of the down- and the up-ailerons. When connecting, or coordinating, springs are used, in order to keep the ball centered in a climbout, you must fly with the controls crossed.

Further uncertainty as to what the rudder is for develops from experience with airplanes that during a left turn require a little bottom rudder to keep the ball centered. This may be caused by the slightly wing-heavy way the airplane trims out, or for other reasons. And, in these same bottom-rudder-left-turn airplanes, top-rudder pressure is needed in right turns to keep the ball centered. A good time to try to answer the question "What's rudder coordination really for?" wouldn't be while offsetting the overbanking tendency in a climbing turn or the unbanking tendency in a descending turn. Certainly any time rudder is required to keep a turn going properly, the customer finds it mighty easy to think of the rudder, at least secretly, as the basic turn control. Even one of the best-selling textbooks ever published once said that in order to turn, you push the rudder the way you want to go. In later editions, this sentence was deleted. We all make mistakes.

Our newly designed airplanes require less and less rudder coordination. In some of the singles, there's little call for rudder coordination other than in the climbout, when it really isn't coordination but yaw control. Even that could be eliminated. In some of the twins with contrarotating props, it has already disappeared.

RUDDER TURNS

Still another area in which the precise function of the rudder can quite logically be misconstrued is rudder-only turns. Cruising along, hands off the stick, you can cause the airplane to bank simply by pushing on one of the rudder pedals. The steepness of the bank can be controlled with the rudder as well. It is also possible (and required for aircraft certification) to come out of a turn with only top rudder, all due to the dihedral effect when the airplane is skidded or slipped. Certainly this nourishes wrong thinking about the proper operational concept of what the rudder's for. Unless a pilot learns to think both consciously and subconsciously that the rudder does not turn the airplane, he is a candidate to spin out of the bottom of a turn onto crosswind or final.

WHAT IS THE RUDDER REALLY FOR?

Possibly the best example of the rudder's intrinsic value is to be found in helicopter flying. Almost anything the helicopter pilot does or thinks causes the nose to want to swing one way or the other. So the pilot learns early that the purpose of the helicopter's "rudder" is to control yaw about the vertical axis of the machine. It can produce yaw if he wants it or prevent it if he doesn't. And that is what it is for in a fixed-wing aircraft, as well. Likewise, the purpose of the elevator is to rotate the airplane about its lateral axis (regardless of ground-referenced "nose-up" or "nose-down"), and the ailerons' purpose is to rotate the airplane about its longitudinal axis, also regardless of ground-referenced attitude, which causes us to think in terms of getting a wing "up" or putting it "down." "Up/down" thinking means we're trying to fly the ground rather than the wing. There's only pitch-up, pitch-down; yaw-right, yaw-left; roll-right, roll-left. The pilot (or the floor of the airplane), not the ground, is the basic reference point in the exercise of three-axis control.

GROUND LOOP

A final irony of misunderstanding the rudder and of the overteaching of rudder/aileron coordination is the experience of the poor fellow who lands a tricycle with drift, and the airplane tips slightly on touchdown and lowers a wing. When he instinctively tries to pick up the wing with ailerons, which is the correct thing to do, the pilot's long teaching of coordination of aileron and rudder movements can cause him to add "top" rudder. Since the rudder is hooked to the nosewheel, this causes a strong swerve *away* from the low wing and increases the tipping tendency of the airplane. Next thing, he's got a wing tip on the runway.

Once the nosewheel is on the ground, coordination needs to be forgotten. It's time for crossed-control operation: left aileron, right rudder; or, right aileron, left rudder, depending on which wing is down. The airplane is then a ground vehicle, and the cure for a "skid" is the same as in a car: steer toward the skid, not away from it. Once the airplane is on the ground, the centrifugal forces the nosewheel can set up in a turn toward a low wing will get the wing up immediately even though this reduces slightly the lifting power of the down-aileron on the low wing and the push-down power of the up-aileron on the high wing.

Originally these misadventures in landing the tricycles with steerable nosewheels were statistically coded as ground loops. But a tricycle gear is stable, and when landed with drift it tends to straighten out rather than ground-loop. Nevertheless in these coordination cases, at least, it looks like a ground loop and the results are the same as in a ground loop in a conventional-gear airplane, or taildragger, where the center of gravity of the airplane is behind the main wheels. The statistical designation finally became "ground loop/swerve." A ground loop and a swerve are not the same thing, but the swerve designation is correct because tricycles don't ground-loop. They swerve because the pilot turns the nosewheel the wrong way with a wing down after touchdown, a victim of his airborne rudder-aileron coordination training.

Let's review a bit of the history of one of the most deserving general-aviation aircraft ever built as a final word on the role of the rudder. The goal of its designer was to produce an airplane easier to learn to fly than anything previously available and which also would get rid of that albatross around general aviation's neck, the stall/spin accident. An understanding of how the designer achieved these goals throws a lot of light on what, in other aircraft, a pilot has to do in order to, you might say, protect him from himself.

The airplane was the Ercoupe, the product of 10 years of research and development by Fred Weick in his spare time as a leading R and D man at the National Advisory Committee for Aeronautics. And it was a beauty: low-wing, wide tricycle gear, twin-tail, two-place, all metal except the fabric wing-covering, an inverted four-cylinder air-cooled engine.

I first flew the prototype with Bob Sanders, an MIT graduate in aeronautical engineering, navy trained, and erstwhile NACA test pilot.

The Ercoupe's nosewheel and ailerons were interconnected—there were no rudder pedals—and steering with the control wheel was quite carlike. After a bit of wandering around the airport and learning that the one pedal on the floor operated brakes on the main wheels simultaneously and was not to be touched on takeoff, we were ready to go.

At about 200 feet on climbout, I was holding what seemed an optimistic enough climb attitude when Bob said, "Pull the wheel all the way back."

"Not me," I answered.

With that he pulled the wheel on his side all the way back. In a high nose-up attitude, the airplane shivered a time or two, the nose went down a few degrees, and then the airplane maintained a steady climb, wheel still all the way back. Bob then rolled it from side to side, still climbing, in and out of banks of at least 60°. Steady as she goes. Finally he leveled the nose and immediately did a steep 180° downwind turn that was on the tight side. No problem.

Our landing was crosswind. My first touchdown wasn't exactly a comfortable experience, but Bob had stressed that the trick was to crab as necessary to keep the airplane moving straight along the runway and to touch down tail-low, still on the crab heading, and then turn the wheel loose. Depending on the crab angle, this might cause the airplane to tip some, but in the process of straightening out, hands-off, the yaw would counter the tilt, keep the wing tip off the ground, and finally pick it up.

If turning the wheel loose on tail-low touchdown proved too much for comfort, turning the wheel slightly *toward* the low wing would accelerate the wing's rise and the swing to runway heading. After which one could start steering. (If this sounds a bit spooky, it was, but Weick was ahead of his day. In autoland, a 747 touches down crabbing and straightens out, à la Ercoupe.)

What were the Ercoupe's benefits? The carlike steering during taxi and in the takeoff run was a great improvement over steering by foot pedals; the tricycle gear eliminated the ground-loop problem so long as the pilot either left the airplane to its own devices after touchdown or steered slightly *toward* a low wing; and the Ercoupe's soft oleo trailing-beam landing gear was designed to handle the side loads imposed in a landing with enough crab to eliminate the drift in a direct 25-MPH crosswind.

On my first flight, we probably would have spun in anything else. We didn't because the Ercoupe did not have quite enough up-elevator travel to reach the stalling angle of attack.

It had enough to make a tail-low, power-off touchdown, but to keep this much travel from causing the airplane to stall with the propeller blowing on the elevator and increasing its effectiveness, as in a climb, the engine had been canted down a bit. With full power, it pulled the nose down as much as the elevator was trying to raise it.

Like the nosewheel, the twin rudders of the Ercoupe were interconnected with the ailerons. From his study of stall/spin accidents Fred knew that as well as moderately restricted elevator up-travel, he would need ailerons that would roll the airplane under all flight conditions. That way, in an attempt to get the wings level coming out of a steep, tight turn a pilot could not, instead, pull the airplane out the bottom of the turn due to adverse yaw from full down-aileron deflection on the low wing and the resulting increased angle of attack of the outer section of the low wing.

Fred accomplished this with differential aileron travel and differential rudder travel of the twin rudders. Turning the control wheel to the left would start the right aileron down slowly and the left one up much faster. Moving the wheel still farther to the left would cause the rudder on the left tail to move out a bit, while the other rudder moved not at all. With full left-wheel deflection, the right aileron would be in trail position and the left one would be up about 60°. At this point the left rudder would be fully deflected to the left, which was enough to keep the ball in the middle in this maximum roll-rate control setting. In effect, Fred got maximum roll rate mostly by pushing one wing down. And with his differential travel of ailerons and rudder, he achieved nearly perfect automatic aileron/rudder coordination. But the most significant thing was that the ailerons always worked, even with the stick full back. That was 40 years ago. We still have airplanes today that you can spin out of a stall, feet on the floor, simply by giving them full opposite aileron when they start to roll off to either side.

Finally, with no direct access to the rudders, the Ercoupe pilot had no way to offset yaw in a climb. This was taken care of by canting the engine slightly to the right.

The Ercoupe couldn't be stalled in a climb or a glide or in a turn, nor could anyone spin the airplane. Some Ercoupe accidents have wound up in an FAA stall/spin/mush category, but actually they were "mush" accidents due to atmospheric effects that resulted in a 10° to 20° nose-down contact with the ground, which made it look as if the aircraft had stalled.

It was easy enough to see what a brilliant concept the Ercoupe was. Certainly it should (and did) turn on a lot of people who found ground-looping taildraggers or spin training not quite their dish. When the Ercoupe went into production in 1939, there were only some 15,000 pilots in the country. It started selling well enough. But in a short time the company had to turn to war-materials production. After the war Ercoupe production was resumed and several thousand were sold as fast as they could be built. But by 1948 the private-owner pipeline had been filled, sales fell off, and ERCO moved into other fields. Sad to say, Fred had a four-place Ercoupe and a small twin on the boards, and the experimental airplanes nearly finished.

Fred Weick—one who dared to be different and to whom aviation progress still owes a lot for his genius in exploring the basics of aircraft docility. The Ercoupe has given us a message we should never forget: we control our wings with elevator and aileron; the rudder is essentially only a glorified trim tab, not a major flight control. But the customers' decision was that they would prefer to protect themselves from the evils of the rudder, which Fred so cleverly banished mechanically and aerodynamically.

ATTITUDE VERSUS ANGLE OF ATTACK

Having come to agreement, or at least a truce, on what makes us go slow, fast, up, down, what the rudder's for and how an airplane turns, it's time now to think about the most fundamental concept in flying: angle of attack, or the angle at which the wing is striking the air.

Almost everyone has some pet illustration of angle of attack. My favorite is visible in water skiing. The high-powered boat races along, and in its wake the skier hangs on to the rope (thrust) for dear life, knees flexed, with the skis planing on the surface of the water at almost zero angle of attack. If the operator of the boat gradually reduces power, the angle of the skis to the water increases, until finally at some low speed and a pitched-up angle, the skis no longer produce enough lift and the skier sinks. In flying, with a comparable angle of attack, we sink, too.

Angle of attack is the heartbeat of the airplane's wing. If the angle gets too high and the airspeed too low, the wing cannot provide the lift needed to sustain the weight it is carrying. And down we come, until the elevator is used to reduce the angle of attack below that at which the wing stalls. Thus, since the perception and control of angle of attack are of primary importance, we need to confine our angle-of-attack ministrations within approximately a 7° range. The maximum usable angle of attack of an airfoil is about 16°, in the average airplane of average aspect ratio, average wing-section, and so on. Beyond that angle, it stalls. Since between 3° and 4° are needed just to support the airplane at cruising speed, at 1 G, that leaves the pilot 10° or perhaps 12° to play with between cruise condition and the slowest practical flight condition, just off the stall. Which isn't a lot to work with.

In our efforts to judge the angle of attack of the wing by the attitude of the fuselage, we tend to forget that the wing is set into the fuselage at a 3° to 4° nose-up angle so that the fuse-lage will be in a level attitude in cruising flight. This is, of course, the angle of incidence, but in our add-on angle-of-attack ministrations, it counts as angle of attack.

If we want to climb at full power and maximum rate of speed, we have to increase the angle of attack of the wing approximately 3° to 4° over what it was in cruise in order to have 1-G lift at the lower speed in the climb. Which means that in a stabilized climb at maximum rate of climb speed the angle of attack is around 6° to 7°.

But this is not where the nose is. In the artificial horizon, it winds up on the 10°-up bar in our average piston airplanes. The reason for this 10° pitch attitude of the fuselage is that our airplanes tend to climb at maximum rate of climb speed up a 6° flight path. In attitude terms, the 10° pitch angle of the fuselage plus 3° angle of incidence of the wing minus the 6° upward flight path gives us our 7° angle of attack. In flying at this angle of attack, we fly at the speed at which we get 1-G lift with the least drag and consequently have the maximum power available for climb. This speed is 1.3 times stall speed and is our safest slow-flight speed as well as most efficient climb or approach speed.

If we want to climb at maximum angle of climb speed, the ice starts getting thinner, and the ante goes way up. The nose has to go still higher, and at the lower speed in this climb regime the angle of attack has to get up to 12° to 13° to provide 1-G lift, and we're operating close to stalling angle of attack.

In our initial training and flight demonstrations, we get off to an unfortunate start in trying to correlate attitude with angle of attack. This kind of erroneous correlation occurs in power stalls. The principal factor in this exaggerated visual-ization of stalling angle of attack in terms of attitude derives from the manner in which the stalling attitude is reached. When the angle of attack of a wing is increased at a steady rate right into the stall, the wing will go to a higher angle of attack, and the nose consequently to a higher attitude before the wing stalls, than if the stall is reached in a lower, constantly held attitude. In the faster entry into the stall, the wing really lets go; sometimes you can find yourself looking 60° down over the nose at the ground, wondering what happened. This, however, can be avoided by starting the nose down just before the stall occurs.

AN EXPERIMENT TO SEE FOR YOURSELF

One way to separate angle of attack from, or better, to correlate it with, attitude, is to stabilize an airplane in climb configuration with full power at maximum rate of climb speed. What we see, as and for the reasons previously indicated, is a 10° nose-up attitude, as we climb at a 7° angle of attack.

Next, gradually reduce power and simultaneously lower the nose gradually until you have the airplane flying level, still at best rate of climb speed. Stabilized in this flight condition, you get the only possible reasonably close look at attitude and angle-of-attack correlation. The nose rides only 3° to 4° above the level-attitude bar in the artificial horizon, which is the 3° to 4° add-on to the angle of incidence which gives you, as in the climb, the desired 7° maximum lift/drag ratio angle of attack of the wing.

Finally, close the throttle and glide, still at the best rate of climb speed, which is close to recommended approach speed. Stabilized, the nose rides down only a few degrees below the horizon, and these degrees in attitude plus the angle of incidence gives us our 7° angle of attack in our glide.

The message is that visually or attitudewise in these

straight-flight conditions, observation of angle of attack is illusory. In all three of the flight conditions described, from nose up 10° in the climb, to only moderately up in level flight, and down only a little in the glide, the angle of attack was the same.

It might be well to repeat the exercise, gear and flaps down, using a 3- to 5-knot lower airspeed. This would show the attitudes needed in this configuration to keep the airplane flying at its most efficient and safest slow-flight angle of attack —namely 7°—in climb, level flight, and power-off glide.

May we have a moment of silence for the poor fellow who, on a hot day with a full load, flies into a downdraft in his climbout and starts settling, and then, thinking that the stick controls altitude, maintains or increases his normal climb attitude. We need angle-of-attack indicators, but they've failed to sell because the concept is abstruse, and the airspeed indicator has seemed an adequate substitute, which it isn't.

STEEP TURNS AND STALLS

The relationship between attitude and angle of attack evaporates rapidly and fails us where we need it most, in a steep turn, which is the preserve of the unintentional stall. In cruising flight, attitude observations still serve us well as we go into a standard-rate turn of about 15° bank. A light backpressure on the stick is necessary to increase the angle of attack enough to provide the additional lift necessary for the wing to lift us around our curved flight path. The speed drops a bit and the airplane wants to loosen the turn as necessary to hold its cruise-trim speed, so we have to keep light backpressure on. The nose rides a bit higher going around the horizon than it did in level flight. The scene is placid. If the nose is carried higher than needed to keep the turn level, we climb. Not enough back pressure held and we lose altitude in the turn.

Even in a 30° bank turn, which is much more lively because of the increase in rate of turn, the too high or too low nose attitude is still small but quite evident, though it is more difficult to find the exact amount of back pressure that will keep the nose at the right place and the turn level.

In a 45° bank the attitude clues are much less evident and much more difficult to manage with precision. Past 45° they simply become too small and too touchy to control easily, and we have to look elsewhere for additional clues as to whether we are carrying a safe margin between us and the stall, or in other words, whether we are flying with more than our desirable maximum 3° to 4° add-on angle of attack.

At this point let's agree that attitude in 15° to 30° banks serves us well as an indicator of angle of attack. With a bit more effort it works well enough between 30° and 45°. And beyond 45° let's go into hold mode before making a final assessment of attitude values in steep turns, until we explore the relationship of angle of attack to another variable: airspeed.

AIRSPEED VERSUS ANGLE OF ATTACK

If you find nebulous the quest for visual clues of angle of attack, or the correlation between attitude and angle of attack, you are not alone. Many pilots think more in terms of airspeed: high airspeed, low angle of attack; low airspeed, high angle of attack. Within limits this serves us well enough, the limits being prescribed by how much airspeed is enough and how much not enough. In any situation, many of us theorize, there's an airspeed, within the operational limits of the airplane, that will keep us a safe number of degrees away from the stall. So "enough" becomes the paramount question in every second the clock ticks during climbout, approach, and landing.

Each owner's manual lists correct indicated airspeed for maximum rate of climb, maximum angle of climb, stall speed with flaps up and down, proper approach speed with flaps up, full down, or in between, all at gross weight on a standard day in smooth air and straight-ahead flight. Corrections for variations in density altitude and, in some manuals, all-up weights are specified. All of which answer the question of how much airspeed is enough reasonably well under the conditions prescribed. In the case of twins flying on one engine we have an "or else" speed for single-engine maximum rate of climb. (I call it "or else" because in most of these airplanes, if you're just a few knots too fast it only flies level, and if a few knots too slow, it loses altitude.)

One of the prescribed conditions for the desirable indicated airspeeds for climb, approach, and stall, which have been mentioned, is straight-ahead flight. Let's now take time to examine the exceptions to that condition.

Mel Gough, a senior test pilot at the NACA facility at Langley Field, hit the nail on the head as never before or since, at least for me, concerning the proper concept of attitude, airspeed, and angle of attack. I wish I could quote him exactly. Essentially, he said that it is possible to stall flying straight up or straight down, and to stall at any airspeed, it all being just a question of angle of attack. That really clarified, for me at least, that neither attitude nor airspeed is to be considered a wholly reliable indicator of angle of attack, even in straight-ahead flight. After all, a snap roll, starting with the proper entry speed, in level straight-ahead flight is simply a horizontal spin at higher-than-normal stall speed.

As with attitude versus angle of attack, our use of indicated airspeed as an additional indicator of our angle of attack runs us afoul of even greater limitations in steep turns. Not only are there instrument errors to deal with, but we are trying to use a rate instrument to get a quantitative measurement on something definable only as an angle—the angle between the relative wind and chord line of the wing.

In our 15° to 30° banked turns, we use a combination of attitude and airspeed information with a cross-check on the altimeter and rate of climb. If, say, we start the turn with cruise trim and cruise power, we become accustomed to a drop of around 10 knots in indicated airspeed in the turn. We also become accustomed to the back stick-pressure needed to keep the turn level, and that tells us something. A larger drop in airspeed is a signal that we may be climbing, so we lighten the back pressure a bit and cross-check the nose position, the altimeter, and rate of climb. Or, we may get the first signal from a glance at the altimeter, next check nose attitude, and finally verify that with an airspeed check before making a stick adjustment. And so on, with similar and opposite corrections for gain in altitude in a turn. We have to use everything available because we are not adequately instrumented or visually and physically equipped to make level turns. Even a change in wind noises or the sound of the engine may require an airspeed/attitude cross-check.

In our turns we get off to a bad start anyhow in many of our airplanes due to pitot/static location errors. As we all know, the pitot tube for the airspeed indicator needs to be pointed straight into the "wind" to give an accurate pressure reading. When the airplane is climbing or in a turn, the tube is striking the "wind" at an angle so there's an error, and the slower we go—that is to say, the higher the angle of attack—the greater the error and the less reliable the airspeed-indicator reading.

What we often overlook is that in a turn the static vent can also be subject to varying changes in pressure in different degrees of bank and airspeeds. This affects the altimeter and the rate-of-climb instrument, which may show a climb during the first part of a turn entry when actually there has been little change in altitude. And later, settled in the turn, they may show a loss of altitude that really only reflects the static vent's becoming again a static vent. Or in some cases, the location of the static vent may produce just the opposite indications.

THE BANK STEEPENS

In banks of 30° to 45°, we get busier with our cross-checks because keeping the turn level is harder to do, especially if there's an indistinct over-the-nose horizon and not even a flat piece of ground to look at as a flagstone for judgment of attitude. In such a case, with a rapid altitude loss, we may relapse into thinking of back-stick as meaning up, and thereby increase angle of attack. At this point, when our airspeed and attitude information no longer tell us enough about angle of attack, we have, fortunately, a sort of angle-of-attack warning indicator, namely G-load. It takes 1.4 G's to maintain altitude in a 45° banked turn. That G-load does not tell us what the angle of attack is, but it does tell us to keep an eye on our airspeed, because an increasing G-load raises the stall speed.

From a 45° bank on, and especially by the time we get to a 60° bank, we're in a sort of never-never land. Except for the pitch-up rate factor going into a stall mentioned earlier, stalling angle of attack is constant but stall speed varies with weight. With the 2 G's it takes to keep a 60° bank turn level, you suddenly have, if you're flying a four-place airplane, the weight of eight passengers aboard, not to mention that the empty weight and the fuel weight have doubled. Think what this weight would do to you in an attempted takeoff and climbout! Then consider what it must do to your stall speed. It's no wonder the steep turn is such a trap. The G-load means a frightful increase in drag and loss of airspeed, and it can also increase the normal upright stall speed reading by as much as 20 to 30 knots. So that, say, 70 knots indicated does not mean the same thing in a steep turn as it does in an approach. G-load does not change the angle of attack at which the airplane stalls, but it jacks up the stall speed enough that it might catch up with a seemingly adequate speed in a turn.

STEEP TURNS ARE EASIER SAID THAN DONE

I hope you're not getting impatient with so much gloom-and-doom talk about the steep turn. I can't in good conscience pass up the opportunity to try to bring it into focus as one of flying's most underrated hurdles. I once compiled some figures indicating that 70 percent of our fatals occurred from loss of control in turning flight at low altitude. The CAA got annoyed with this and employed a hotshot independent statistician to study their files: the official figure became 63.4 percent.

Pilots worry a lot about thunderstorms and icing and severe weather and engine failure, but loss of control in turning flight at low altitude (where it always seems to happen) is our greatest single producer of accidents, and nobody worries about it. This may be because our turn stall demonstrations are little more convincing than our straight-ahead power stalls as far as angle of attack is concerned.

One of the most significant figures that ever came out of FAA's statistical department resulted from a sorting of spin-in files according to the rate-of-climb capability of each airplane. The airplanes with the high rates of climb had the fewest spin-ins per hundred aircraft.

Which figures. To start at the bottom of the totem pole historically, pilots flying 36-HP C-3 Aeroncas and 37-HP Cubs used to spin after little more than 360° of steep turn. They got into a power stall and spun in, power on. Those airplanes, with a good engine/prop matching, would steam along at a lusty 60 MPH at low altitude; their zero-bank stall speed was around 40. In virtually all airplanes, pulling 2 G's, as you would in a 60° bank turn, raises the stall speed 25 knots. In these early models 40 plus 25 equals 65, and you hit the jackpot.

Moving forward 30 years or more, let's try to get a slow-motion look at a stall in a turn, power on, in a retractable which has a 700- to 800-FPM rate of climb even at cruise power. It has a good speed ratio—that is, from maximum legal zero-degree bank stall speed of 69 MPH (or less in some cases) you may add 100 MPH for cruise speed.

In a 45° bank with cruise power in these clean aircraft, indicated airspeed may drop as much as 20 MPH, but the airplanes will fly round and round all day at constant altitude with 1.4 G-load. So where's the stall demon lurking? You can't really demonstrate a power-on stall in these airplanes unless you either increase the bank well past 45° and keep turning until the indicated airspeed comes down to meet the G-load elevated stall speed, or use a low entry speed.

The latter method relates to spins that occur after partial or complete power loss and a desperate attempted turn-back on takeoff, and after overtightened and overbanked turns into final. Somehow these catastrophes suggest that in practicing steep turns and trying to acquire skill in keeping them level we tend to think that a steep turn is not a proper steep turn without its in-practice associated high G-load. Forgetting that we just haven't far to go in stalling angle of attack with a low entry speed.

If keeping our turns level reduces our airspeed, sometimes to a number perilously near the stall, we might ask who says turns must always be level. In instrument flying, level turns are a requirement, but we are told to keep our bank shallow —the lower we are the more so—and at the first sign of an increase in airspeed and loss of altitude, we shallow the bank. In VFR operation we do not do that but instead tend to increase the G-load in a steep turn. And thus we move into a critical flight regime.

Why not just avoid steep turns? In our weather flying, we are learning now what the airlines started learning years ago —we—and they—do not fly into severe weather. Our progress, and safety, have come through avoidance of severe weather. Knowing the trouble they cause, then, why can't we simply avoid turns steeper than a 45° bank? And even in those, if they are to be prolonged, why don't we bring the power up for an extra margin of safety?

In a 60° bank, no matter what you're flying, if you don't put on 2 G's, you'll quickly be looking at a 1500-FPM or more descent. If the nose gets low enough, and you keep pulling on the stick instead of reducing the bank, you can find yourself in a rapidly tightening spiral and go right past the redline on the airspeed indicator.

WEIGHTS AND MEASURES AND TURNS

Let's look at the G-loads required to make level turns of varying degree of bank (provided you have the power to pull the airplane around the track fast enough to keep flying at those weights).

		Degree of Bank			
	30°	*45°*	*60°*	*70°*	*80°*	*90°*
G-load	1.2	1.4	2.0	3.0	6.0	Infinity

Don't overlook that last word on the G-load line. We've all seen pilots do vertical banks without the wings coming off. The CAA explained once that they get away with it by carrying the fuselage at a slight angle of attack, getting just enough lift to keep the airplane stuck on the wall of the silo as it goes around. It goes without saying that to continue flying with an infinite G-load in a vertical bank, the speed would also have to be infinite. So we can back up from there, for a rough indicated airspeed yardstick.

An 80° banked level turn? Note the 6-G requirement. Our wings are certificated in normal category operation to fail under that load. So, no sale.

Look at the difference between 60° and 70°. Ten degrees is not a big difference in bank, but the G-load goes up 50 percent. Even holding a miserably uncomfortable 3 G's, we'd do well to remember that our normal-category airplanes, at gross weight, have a 3.8-G limit load factor. Why stand on the edge of a cliff and see how far it is possible to lean backward without losing your balance?

So, if a 60° bank, which gives a high rate of turn, is a safe operational limit, what indicated airspeed, power on, is needed to stay safely above stalling angle of attack? There are no published figures on this. All we have to look at are indicated stall airspeeds in various angles of bank, *power off*. I think the reason for the power-off prescription is that at the higher angles of attack, indicated airspeeds are less and less reliable, and that in a power-on turn we go to higher angle of attack before stall than with power off. Also, it is easier to overdo the G thing power on. Power-on stall speed? Certainly an engineer would be the first to ask, "How much power?" Pilots think in terms of judicious approximation, but engineers have to think in terms of the difference between .001 and .002. So they eliminate the power-on variable and give us power-off figures, which they can at least teeter on. The power-off figures give us a vital message as to how much stall speeds can vary with G-load. Power on or power off.

In the accompanying table, so as to emphasize the increase in stall speed with G-load, the 0° to 60° part of the G-load table will be repeated and below it will be shown below 0° the stall speeds shown in owner's manuals. The stall speeds are KIAS, except those for the Warrior, which are KCAS; the increases are in knots, the airplanes are at gross weight, forward center of gravity, flaps up, power off. First off, it would be reasonable

		Degree of Bank			
	0°	30°	45°	60°	60°
G-load	1.0	1.2	1.4	2.0	2.0
	Stall	Increase in Stall			Stall
Cessna 152-2, KIAS	54	4	10	23	77
PA-28-151 Warrior, KCAS	56	5	10	13	78
Cutlass, KIAS	51	4	10	21	72
PA-32-300 Cherokee Six, KIAS	54	4	10	23	77
PA-31 Navajo, KIAS	76	5	10	23	99

to ask why the 70° to 90° part of the G-load table was dropped. I think the reason those stall speeds are not available for our general-aviation piston airplanes is that they would exceed the highest speed we could produce in these turns with the power available. Thus a stall would be inevitable.

Otherwise, if the indicated stall speeds go up, as the table shows, 20 to 25 knots in a power-off, level 60° bank turn with 2 G's, what indicated should we carry power-on in order to have a safe margin above stalling angle of attack? In our approaches we use a safety-valve speed of 1.3 times stall speed. Assuming that power-on figures wouldn't be materially different from those in the tables, we could use our 1.3 standby.

Let's use the Cutlass figures as an example. Its 51-knot stall plus a 21-knot increase would give us 72 knots to add .3 to, or 94 knots. In short, we could look at the figure at the bottom of the green on the airspeed indicator, add 21, and multiply the result by 1.3.

Or we could go fly and see what the indicated airspeed is in a stall in a 60° banked level turn in a Cessna Cutlass. Answer: it can't be stalled solo in a level 60° banked turn with as much as 23 inches and 2300 RPM (about 60° perent power). In such a turn at 5000 feet MSL, just before I went bye-bye after several 360's, holding the turn level as I could, the indicated got down to only 90 knots. So I was a bit below the 1.3 computed "proper" speed in the turn but still well above the power-off stall speed of 72 knots. With four souls aboard, a bit higher power setting would have no doubt kept the indicated at 90 knots.

There's a very interesting and commendable thing in the Warrior manual. They classify a steep turn as an acrobatic maneuver, along with lazy eights and chandelles, and give 108 knots (an average low-altitude cruise speed) as the recommended entry speed. These maneuvers are approved only with the airplane at or under its utility category gross weight of 1950 pounds. In other words, 400 pounds light. At this weight I estimate the Warrior's stall speed in a 2-G 60° banked turn as 71 knots, and one and three-tenths times this gives us 92 knots as the minimum speed we'd want to be able to maintain in the turn so as to keep our angle of attack at our faithful 7°. Starting with an 108-knot entry speed we'd have a margin before slowing too much.

At the Warrior's normal category gross of 2350 pounds, the stall figure in a 60° banked 2-G turn is 78 knots, or 102 is our 1.3 figure. Which we probably wouldn't be long getting down to with a 108 entry speed.

The significant thing about these entry figures is what they suggest about the sequence of events with an entry speed of, say, no more than maximum rate of climb speed.

In the Warrior at 2350 gross the maximum rate of climb speed is 87 knots, or only 9 knots above the 60° bank stall speed. I can't miss the opportunity to visualize the fellow with an engine failure on takeoff. If he gets the nose down enough to give him an entry speed of 87 knots, once he has the airplane in a steep turn, from habit as well as his hurry to get turned back, he is going to put on the G-load normally associated with a 60° banked turn entered with cruise speed. To get 2 G with only 87 knots airspeed will call for at least a 15° angle of attack, so it's no wonder that in the first second or so of the frenzied turn-back attempt the airplane spins. And, of course, it doesn't take an engine failure. It can happen in a steep, tight turn onto base or final.

I hope this illustration will not be taken as a reflection on the Warrior. The scenario repeats time and again in everything we fly into a steep, tight turn with a low entry speed. There's just no way to pull 2 G in a low-speed steep turn without going to stalling angle of attack.

Finally it doesn't seem proper to go this far into what indicated airspeed and G-load can tell us about how well we're doing on angle of attack without mentioning that indicated airspeed is, on occasion, used for attitude information.

This most commonly happens when a VFR pilot has lost visual reference or an instrument pilot has either lost the vacuum pump or the artificial-horizon instrument.

The airspeed/attitude concept starts with the airplane flying straight and level at normal cruise power indicating, say, 150 knots, and the pilot flying blind. If the speed, straight ahead and wings level, drops to 140, that means the nose is above the horizon, or above cruise attitude. A light momentary forward pressure on the stick gets the speed back to 150. With 160 indicated, the nose has to be low, so a momentary back stick-pressure corrects that. So long as the turn needle is kept straight up with the rudder and the ball centered with the ailerons, this works well enough.

The tough part of the system develops when the pilot's problem is how to get from a high indicated or a low indicated airspeed back to the 150, which he associates with the serenity of straight and level flight.

Say that from a diving turn or spiral he has been able to get the turn needle straight up and the ball centered and is therefore in a straight-ahead dive, indicated airspeed 200 knots, altimeter unwinding madly.

Naturally the pressure is now on to get out of the dive. Too much back pressure on the stick, though, and the aircraft structure could be overstressed, or the airplane could stall and might spin. Just enough back pressure to get the airspeed

---

to begin coming down is the first goal, granted that the airplane is kept out of a turn.

But if the back pressure is kept on until the indicated airspeed is down to 150, the airplane will be headed approximately straight up, because of the lag in the attitude indication of the airspeed indicator. This lag in attitude indication explains the accounts of witnesses who hear an airplane in a screaming dive, prop snarling, then hear the sound diminishing and receding, followed by a laboring sound from the engine (at the top of the zoom).

The trick of the trade is to fly the *trend* of the airspeed needle. When the airplane starts coming out of the dive as indicated by the start of a *decrease* in the indicated airspeed, it's time to reduce by a bit the back pressure on the stick. Soon after that, it is necessary to hold a bit of forward pressure to slow the rate at which the needle is moving toward 150. Do it right, and it's almost like a three-point landing: you'll ease off the forward pressure just in time to stop the airspeed needle on the cruise speed and you'll be in level flight.

Conversely, if, starting from a stall-horn, low-airspeed reading, forward pressure, even though light, is held on the stick and the airplane is kept out of a turn until the indicated is 150, the airplane will be pointed nearly straight down or even past a vertical attitude. The trick again: fly the *trend* of the airspeed reading. As soon as the speed starts increasing, start applying back stick-pressure to slow the rate of increase, and you'll come out close to level at 150.

The rule: back pressure when the speed is increasing; forward pressure when it's decreasing. Which may seem backward, but it isn't. This all comes easier to some than others (I was an other).

TRIM VERSUS ANGLE OF ATTACK

Angle of attack is not controlled by airspeed. It is purely geometric. It is the angle at which a tilted wing strikes the "wind" as it moves through the air, and this angle is controlled by the position of the elevator, which, in turn, can be controlled by the trim-tab setting. Since we do not have angle-of-attack indicators in most general-aviation aircraft, it might be better to put the subject in pilot vernacular and consider a more everyday and hence more proper title for this section: rather than "Trim versus Angle of Attack" let us call it "Trim versus Airspeed," since we think of our trim control as a mechanism for holding a selected speed.

In that sense the trim-tab setting determines the speed at which the airplane wants to fly, to quote from the scriptures according to Langewiesche. This does not tell us what the angle of attack is, but any stick pressure, back or forward, that overrides the trim airspeed and makes the airplane fly slower or faster gives us, in most situations, still another valuable angle-of-attack clue. At least, it tells when and how much we may be straying from an airspeed or angle-of-attack goal we've set with the trim tab.

For instance, on an approach if the trim tab is set to give a 1.3-times-stall-speed reading of, say, 70, with no or only a light back-pressure on the stick, we do not have even to look at the airspeed indicator to know that we've gone below that desired approach speed if we have increased the back pressure. What better message could one have of a deviation from a desired angle of attack?

The situation is the same for a proper climb speed (or angle of attack). The airplane wants to climb at the airspeed the trim tab is set for. If we override this with extra back pressure, we know we've used up a part of our airspeed or angle-of-attack safety margin.

So anytime we're overriding a trimmed airspeed, we need to keep in mind the airspeed at which the airplane wants to fly. At least that will tell us if we're tending toward operation at the lower or upper part of our small 7° angle-of-attack cage.

But, like all the rest of the seemingly pat answers in this flying business, this one also has its limitations. In a balked landing, for instance, with some extra pitch-up from full power or in a steep turn with approach or takeoff trim, we have to remember the trim-speed setting and take charge. Isn't it striking that so many of the answers to angle-of-attack problems are to reduce angle of attack, with forward pressure on the stick?

There are variations in the use of stick-pressure override as an angle-of-attack indicator. For instance, many a person has learned to fly a Cessna 150 with the trim tab always set for cruise speed. This means that the airplane is nose heavy in the climb, but only moderately so. These pilots learn to associate a certain back pressure with the maximum rate of climb speed. And, since the trim of the airplane is not much affected by flap position or changes in power, they also learn that an only slightly different back pressure on the wheel produces the recommended approach speed. Since it produces what you might call a couple of constants, this system is better than having to deal with the variables introduced by different trim settings on every circuit in training.

On the other hand, when these pilots leave the side-by-side trainers and get to retractables, and the faster airplanes, and those in which putting the gear or the flaps up or down causes a big trim change, they have to trim for the speed at which they want to fly, simply by holding the nose as necessary to get stabilized on the speed they want and then trimming the load off the wheel.

There's need to mention a misuse of trim-tab setting ex-acerbated today by the increasing appearance of electric trim-tab switches on the control wheel under the pilot's left thumb. This convenience and great aid in a bad out-of-trim situation is all too often being used as not a trim-tab control but as a primary flight control.

Pilots often fly cross-country without ever taking their thumbs off the electric trim switch or letting it rest for more than a few seconds. Whether they're thinking of controlling attitude with trim or using it to maintain altitude, they are really trying to fly the airplane with the trim tab instead of the elevator. They never get to quite where they want to be because every time the trim setting is changed, the air-plane's stability moments start trying to achieve the appro-priate airspeed for the new trim setting, by either nosing up or down. Even if that's in the direction of the pilot's desires, when an altitude is regained, the airplane goes right through it unless the pilot retrims. If you try to fly the air-plane with the trim tab, the airplane chases a never-ending succession of speeds at which it tries to stabilize. It's better to get the trim set for cruise and then ignore minor excur-sions in attitude or altitude, or, if they're too much, override the trim setting momentarily. When the disturbance has passed, releasing the stick pressure returns primary control to the airplane's longitudinal stability and the trim setting.

AHEAD OF HIS TIME

The best combination of trim/airspeed/longitudinal stabil-ity I ever saw was in a Cessna 172, circa 1955, in which a fellow named Woody Hunt showed up one day at Trenton Airport in New Jersey. Woody was an engineer at Tactair, Incorporated, builders of a unique pneumatic autopilot system using bellows for servos. He had called to say he had something he wanted to show me before they put it to sleep since they hadn't found a buyer.

He had a small black box aboard into which they'd run the dynamic and static pressure lines of the airspeed indicator. The box also included a sensor for change in rate of change in airspeed, and a valve with a line out the back of the box to their autopilot's elevator-control bellows. The purpose of the rig was to make the airplane fly at whatever speed the trim tab was set for, power on or off, flaps up or down. Or in other words, at a constant angle of attack, which the pilot could set with the trim tab and forget about. In quantity they had projected a price of $200 net to aircraft manufacturers.

The demonstration was spectacular. With the modified Tactair autopilot on, which meant the heading set on the directional gyro would be held, an increase in power as you flew along at cruise would give a moderate climb at cruise speed. Or if power were reduced below cruising power, the airplane would descend at cruising speed. With the throttle closed, it would still hold the trim speed.

If you wanted best climb you'd simply push the throttle in and move the trim wheel until the indicated airspeed was on best rate of climb speed, and it would hold the speed. For an approach you reduced power and moved the trim wheel until the airspeed needle was on the desired airspeed—from there on, the throttle could be used to control rate of descent and the airspeed wouldn't change. In these maneuvers putting the flaps down or raising them would change only the attitude, not the airspeed.

In short, here was an airplane that would fly at a constant

angle of attack; the only pilot skill needed to keep that at a safe figure would be that of moving the trim wheel to get the airspeed needle on the proper speed for climb, cruise, or approach. In turbulence the response of the device to undulations in the atmosphere would cause a slight change in nose position, but no change in airspeed, and the bellows could be overridden at any time with only a moderate pressure on the elevator control. Somehow I feel that the Tactair constant-airspeed device may simply have been born too soon and will not forever remain in limbo.

ATTITUDE FLYING

While considered the modern system, attitude flying is not exactly new. From way back many a pilot has observed that such and such a nose-up attitude was as high as one should go in a full-power climb, and that with the nose down a bit below the horizon the airplane sure wasn't going to stall in an approach.

I think the stress the FAA puts on attitude flying—Attitude plus Power equals Performance—filters through to them primarily via the former military pilots in their employ. It is undoubtedly an effective system when one is flying an airplane with so much power that it will go just about anywhere it's pointed. After all, given enough thrust there's no need of a wing at all, at least to go.

In our piston singles and twins, we do not have the 2000- to 5000-FPM rates of climb in which an 800-FPM downdraft would hardly be noticed. In many of our aircraft that is more than our maximum rate of climb, so we would have to forget about attitude and concentrate on our angle of attack. We are also in a different realm of flying in our approaches, when gust effects can subtract 20 percent of our approach speed as compared with only 10 percent or less for a pilot whose approach speed may be 170 MPH or more. We have to forget attitude at these times and look out for ourselves.

I got a good illustration of this a few years ago. I was on a cross-country trip in a 250 Comanche when a cadet on leave from the Air Force Academy hitched a ride with me to a major airline terminal up the road. He had received his primary training in a twin-jet Cessna T-33 and had never flown anything other than jets. Confused by the Comanche's manifold-pressure gauge, tachometer, and constant-speed propeller, he finally asked, "What would happen if you gave it full power and pulled straight up?"

"You'd come straight down," was my answer.

But this is not to say that attitude flying, or at least parts of the concept, are not of great importance to us. For instance, attitude flying holds the secret of precise speed control. Many's the time you've seen a pilot chasing the airspeed needle all the way in on final. He never gets it settled down and rides the roller coaster all the way in. He may not even know about flying the trend of the airspeed needle.

But he doesn't need to. The easiest way to control the airspeed precisely in a climbout or an approach is simply to put the nose at an accustomed or estimated proper attitude for the occasion, hold it there firmly for a moment, and then glance at the airspeed indicator. If the speed is a bit lower than desired or a bit higher, make a small change to a new nose-lower or nose-higher attitude and *hold* it there. It doesn't take more than one or two adjustments of attitude to capture the speed you want to use. The payoff for precise speed control in an approach is that it makes the time sequence from flare to touchdown uniform under most conditions, which is a big help in the quest for soft touchdowns.

My most serious reservation about the FAA's request that people believe that attitude flying provides all the answers comes from their development of the idea some years back that it is possible to recover from a "departure" stall without loss of altitude simply by putting the nose in a level flight attitude and pouring on the coal. Although they now permit the nose to be lowered a little below level in such a recovery, most of the stress is on adding power. The trouble with this system is that what they're talking about is not recovery from a full stall but recovery from slow flight or at worst an incipient stall. What happens to the pilot who comes awake slowly, and remembering the attitude admonition about getting the nose to a level attitude, finds it not only below the horizon but dropping farther after a real stall? Pull the nose up? He can't because he's already used up all his up-elevator control force. But he tries. And spins.

The "get it level and gun it" concept is even worse in a stall in a turn. In such a stall they tell us to get the wing up first, then the nose, and then gun it. This is talking in terms of where we want to be instead of how to get there. With roll control as well as pitch control lost, trying to get the wing up can only compound the felony considering most ailerons' adverse yaw characteristics. What must come *first* in getting out of this trap is undoing the thing that got us into it, namely a too high angle of attack. The instant cure for that is reduced back-pressure on the stick; after a second or so this reestablishes both roll and pitch control. Then, and only then, is it possible to start flying the airplane again.

Yes, Attitude plus Power does equal Performance, but it is not always the performance the general aviation pilot wants.

2

ATMOSPHERIC EFFECTS ON ANGLE OF ATTACK AND AIRSPEED

Although it is possible to get glimpses of angle of attack at times, and at other times to experience a feel for change in angle of attack, and even to control angle of attack with the elevator, and derivative airspeed with the elevator and throttle, it is important to understand that changes in angle of attack and also airspeed can occur without any input from the pilot. That's why flying is not quite as simple as it sounds or looks.

From the time we light the fuse for takeoff and rotate, until we're back on the ground with the throttle closed, we are hostage to proper control of angle of attack. As a measure of true flying skill, angle of attack needs to be limited, in normal operations, to 7°—which is the angle of incidence of the wing plus our 3° to 4° add-on for best rate of climb, constant-altitude slow flight, and proper approach speed. This isn't much of an operating range, but it's a necessary limitation if we are to keep out of reach of the abrupt changes in angle of attack that can be caused by atmospheric effects. Our 7° limit gives us a needed shield of approximately that same amount below stalling angle of attack.

Angle of attack? It may be well to belabor the point, by

reviewing our earlier analogy between an aircraft's wing and a skier on water. It is easy enough to see how the skier sinks if towed too slowly and that this is comparable to our flying situation when we fly too slowly, straight ahead. But this is about where the analogy ends, for in the air, the invisible surface on which we ski can be tilted upward, downward, laterally, and in combination can give us not only turns but climbing and descending turns and spirals—provided, no matter which way we're going, we do not let the wing reach stalling angle of attack. Additionally, we ski through rapids and shoal waters. In our aerial skiing, the pilot, rather than the driver of the boat, has control of thrust and, in addition, angle of attack, airspeed, attitude, altitude, and heading. If being a free soul appeals to you, this is it. For many it makes flying a lifetime romance.

RELATIVE WIND

Textbook illustrations of angle of attack presuppose sterile conditions: namely, smooth air. Angle of attack is shown as the angle between the mean chord line of the wing and the flight path of the airplane. This line, behind the airplane, can be level in constant-altitude flight, or tilted upward in a climb, or downward in a descent. A continuation of the flight-path line ahead of the airplane is usually labeled relative wind, with an arrow on it pointing toward the airplane. Which is where the relative wind is from if the airplane is flying in a stable air mass, and which is where it isn't from when a pilot flies into an updraft, downdraft, gust, wind shift, or slower-moving layer of air.

In pilot terms we need to think of relative wind not as the angle between the wing and the flight path behind us, but as the angle at which the wing collides from second to second with the air ahead of it. When that air has a vertical or horizontal component of motion compared to the air in which the aircraft has just been flying, it will change the angle of the relative wind and thus change the angle of attack of the wing.

The word "wind" in the phrase relative wind is, in a way, unfortunate. We tend to think of wind in ground-observer terms. An air mass moving across the country produces a wind, on the ground. But an airplane in flight knows no difference between an air mass that is not moving and a stable air mass moving at a steady rate. So in that sense it doesn't really fly in wind at all. It makes its own wind with its movement through the air. But let it fly into an area of air movement— an acceleration or deceleration within the air mass in which it is flying, such as the puff of a gust, or an up- or downdraft —and the pilot is handed a momentary change in angle of attack and airspeed for which he may need to correct. So, conceptually, the textbook illustration is correct for calm air or steady wind conditions, but it doesn't tell the pilot the facts of life he will encounter on most of his flights.

GRADIENT WIND

In seeking an understanding of the dynamics of the atmosphere as they affect angle of attack and airspeed, let's start with one of the simplest and most frequently experienced conditions in approaches: a gradient wind. By definition this means that the wind is coming straight along the runway, but during the approach we descend into a layer of air of decreasing velocity. Say at 100 feet the headwind is 15 MPH and just as we cross the fence with 50 feet, it's only 5 MPH. For the wing this means an abrupt drop of 10 MPH in approach speed, and it is what gives us that sinking feeling at the last in so many of our approaches.

As we descend into slower-moving layers of air, the wing loses lift momentarily until the airplane accelerates to its intended approach speed in the slower-moving layer. The increase in sink rate makes us feel light on the seat and causes us to gun it—sometimes too much and sometimes too little. I doubt that any instructor has ever been willing to solo a student without having seen him gun it a few times in one of those "somebody must have left off a manhole cover" situations.

Even so, failure to add power soon enough, or at all, and even to pitch up at the very last, accounts for many a washed-out landing gear, bent prop, and sprung crankshaft. In these situations the pilot was doing all right angle-of-attack-wise until nature pulled the rug out from under him with that abrupt 10-MPH drop in airspeed. The drop in airspeed may not have used up more than half the margin over stall speed and stall angle of attack, so there was time to correct the situation, but the pilot was too slow to get going with a momentary increase in power. Not necessarily a big increase, but a vital one. Or, if power wasn't available, he may have failed to make a quick-enough deal on trading at least a little altitude for less drag and more airspeed.

I had flown 20 years before I ever heard the phrase gradient wind—20 years of mostly cross-country flying, with a few hundred hours in student instruction and quite a bit more on the right side learning from watching prospective aircraft purchasers fly airplanes they'd never flown before (with a few hints, as appropriate, about desirable positions of the nose above and below the horizon in the process of getting it up and down and on the ground).

The occasion for my formal enlightenment was a private-flying crystal-ball-gazing affair right after World War II at Wayne State University in Detroit. I ran into Fred Weick in the lobby of the hotel just before the evening session. He had come on too short notice to get a reservation, and there were no rooms available. No problem. I had a spare bed in my room to which he was welcome. I'm afraid he paid dearly for it.

We turned in about 11:00 P.M. I should have kept quiet, but the man across the room not only had all the answers about why an airplane flies, but he could talk about the whys and wherefores in pilot's language, in terms of what can be done about them. As it turned out, being a teacher at heart as well as in fact, Fred responds generously to questions from anyone who is eager to learn.

Fred's lecture on gradient wind went something like this: on a nice smooth-air day have you ever, flying along at altitude, closed the throttle, put the flaps down, and glided for a while with the stick finally almost all the way back, wondering if this might not be the way to handle a dead-stick landing if you could find a bushy place to come down in? Those who have tried it get along fine until the last 50 to 100 feet or so. Then the airplane noses down, with the stick still back, and dives into the ground, because it's glided at minimum speed into a slower-moving layer of air close to the ground. Gradient-wind effect.

In working with his W-1, the prototype of the Ercoupe, Fred decided to find out how much oleo travel his landing gear would need in order, in an emergency, to make a landing as aforementioned, granted enough elevator up-travel restriction to keep out of a stall.

In order to get some numbers on gradient-wind effects, he rented an autogyro. Normally they come down, power almost off, steeply, at around 800 to 900 FPM and have enough kinetic energy in the rotor system to stop this descent in the flare. Fred figured that his W-1 with a slightly lower stick-back rate of descent could survive, or at least the souls aboard could survive, a stick-back descent and touchdown if it had 18 inches of travel in the landing-gear oleos.

As it turned out, on days when there was a gradient wind on final, and especially if there were also even mild gusts, just as he was reaching a peak of optimism, the helicopter's 900-FPM rate of descent would abruptly increase to 1800 FPM. Impact velocity varying as the square of the speed, that would call for 6 feet of oleo travel in his small airplane's landing gear. Which was not exactly *quod erat demonstrandum.* Or QED, as the engineers sign off. Which Fred did about 2:00 A.M. when, beginning to feel a bit like an aeronautical ass, I quit asking fool's questions. Almost instantly, it seemed, the phone rang: the 4:00 A.M. call for his flight back to Washington.

Incidentally, the autogyro pilot did not crack up on any of those tests. Just as the sink rate doubled, he'd pour on the coal and get the situation back in hand. Just as we can do in our airplanes if we act soon enough when we encounter the sinking effect of a gradient wind on final.

WIND SHEAR'S THE WORD

Gradient wind is not talked about anymore. The up-to-date and more inclusive phrase is wind shear, which is defined as a change in wind direction or velocity. In short, descending from a headwind component into a layer of air moving directly from one side to the other has the same effect as going into a layer with zero headwind component.

Since the wind-shear concept is so important, I'd like to add, in case the gradient-wind/wind-layer analogy hasn't come through clearly, that the visualization of wind effects is not as important as coming up with the right answer at the right time.

For instance, you might be more comfortable thinking in

terms of an airplane flying headed into a wind with zero ground speed. Say you were flying thus in a free-flight wind tunnel. If someone cut off the fan, you'd be on the bottom of the tunnel pretty quickly. Or if the fan were speeded up, the increased wind velocity would create more lift at the angle of attack being flown and the airplane would go up to the top of the tunnel unless angle of attack were reduced to provide level flight at the new, higher airspeed.

So, if thinking in terms of wind relative to the airplane sustains you, that's as good as any other visualization, so long as it explains that what we have to hang on to mentally is that wind per se doesn't mean anything and the key and vital thing is *change* in wind direction or velocity. It is these changes that put a nonpilot input into where the airplane is going to go or what it's going to do. Which puts upon us responsibility for a prompt angle-of-attack or airspeed correction.

IN THE BIG LEAGUE

Development and perfection of wind-shear measuring equipment is now a major safety project of the government at several major airline terminals, and eventually ATIS will include significant wind-shear information as determined by wind direction measurements around the perimeter of the airport and by laser and other equipment that makes it possible to "see" the air. They are even experimenting with DME equipment to see if possibly the earliest and most useful indicator of an incipient wind-shear encounter may not be an abrupt change in ground speed on final approach. You may recall some of the recent airline disasters that were attributed to wind shear. At Kennedy the airplane landed short, in the approach-light thicket. At Philadelphia the airplane landed harder than the gear could take but on the airport (the pilot was faulted for not trying a go-around at the last, but I think his judgment that he shouldn't risk a spin-in—after all he was in landing configuration with everything hanging out—and 100-percent casualties instead of 10 percent or fewer was laudable). At Denver the pilot unknowingly climbed into an overrunning tailwind at low altitude and landed a mile or so beyond the end of the runway.

Warned of wind-shear conditions, a pilot will be able to carry extra approach speed on final, or ask for a runway change, or start a go-around earlier, or go somewhere else.

In general aviation, unwarned as we will be for no doubt a long time to come, we can be more sensitive to changes in rate of descent and add power more promptly, and maybe even trade what altitude we can spare for additional airspeed and lower angle of attack. As we go through a wind shear, our airspeed instantly decreases, which means less lift for our given angle of attack, and our rate of descent increases. This in turn makes the flight path (behind the airplane) turn down, thus increasing angle of attack and drag. If the pilot does nothing, and has altitude enough, the power in use in the approach will gradually accelerate the airplane back to the relative airspeed it had before entering the slower-moving layer of air. But this is not something to wait around on too long. Best of all, we can be wary of risking landings in the hotbed of wind-shear encounters, that is, in the vicinity of thunderstorms.

May I have your indulgence as I recount a highly prized example of a gust-spiced gradient wind encounter?

Ruuumph. In my transatlantic flying experience (riding right behind TWA Captain Robert N. Buck, with CAB cockpit authorization, in the FAA's line checker's seat in the cockpit) I've seen Bob lay Connies, 707's, and 747's on so softly that it was not apparent that the airplane was on the ground until he started lowering the nose in the roll-out.

One night we got to talking about wind shear and gust effects, and he said that adding half the reported gust increase over the prevailing wind to the bug speed calculated for each approach was all they had and was basically a good system. Then he said, "An odd thing happened the other night. Seven years ago on my first solo trip as a 707 captain we were going into Leonardo da Vinci airport at Rome and were allowing 10 knots over bug speed for gusts. The approach was nice and stable and the copilot commented, 'Look, we're going to land right on the numbers.' "

"We did. Man, was it a hard landing. Shook the airport. I even had the airplane inspected.

"Well, night before last going into Rome on the same runway and under the same conditions the same thing happened again. As chance would have it, there was a stewardess on board who was on duty that first flight seven years ago, and when she got off she said, 'Captain, you haven't improved any.' "

GRADIENT WIND IN CLIMBOUT

This is an atmospheric effect that you might say is on the plus side, since it increases airspeed without affecting angle of attack, although, through the exuberance it generates, it often causes *pilots* to increase angle of attack.

Say you're taking off into a 10-MPH wind and are stabilized at maximum rate of climb speed. The ground drops away far more rapidly than usual, and the rate-of-climb indicator soon shows, instead of its usual 800 FPM, 1000 or even 1200 FPM up. At retail that is an expensive rate of climb, and it is exhilarating.

A common reaction to such performance is to pull the nose a little higher, to maximum angle of climb speed or even a bit higher than that, and this pays a further dividend in visual effects. Unfortunately doing this may whisper to the pilot that he's the cause of this spectacular performance, that the secret of attaining it is simply really to pull back on that stick.

In this lively climb the airspeed might be expected to drop off some, but quite often it doesn't, presumably because as the altitude increases the headwind increases. If the gradient-wind layer is thick enough, the rate of climb may go to 1500 FPM, or almost double the usual amount. This is because in climbing in a gradient-wind condition it is possible to get a loan of kinetic energy from the increased velocity as you go up. The results are the same as would be possible with a very substantial increase in horsepower. But this performance shouldn't be allowed to set a mental trap such as the supposition that back-stick really does mean up or that the power stall attitude must surely be a lot higher than people realize. On a nongradient-wind day, trying to climb in a gradient-wind attitude could invite real trouble.

Otherwise, climbing into a wind layer that has a strong gradient is one on the house, or on some days a couple.

CLIMBING INTO A TAILWIND

So far we've considered gradient wind as it is encountered in approaches and climbouts. We've talked about lower-velocity wind as we come down, higher as we go up. Head on. This time it's a slower gradient we're ascending into.

It may be calm on the runway or the takeoff may be into a light headwind, but off and into the climb all of a sudden a check shows that the power is all right, the airspeed correct, but objects ahead are not moving down in our field of vision and the rate-of-climb indicator may linger on a sickly zero. For some reason the airplane just doesn't want to climb.

It has happened to all of us, especially in summertime and at gross weight, or in high country when we're a bit short on power anyhow unless turbocharged. There may have been a 180° wind shift, and the airplane has climbed into the bottom of an overrunning tailwind. During the period it would take for the airplane to accelerate to its intended climb airspeed, in the new ambient air condition it quit climbing.

Or, temporary disappearance of normal performance could be caused by a wind-shear effect, that is, a 90° change in wind direction. But the acceleration period wouldn't be as long as with a 180° wind shift in the climbout.

In either case we are reminded again that with a given power and optimum climb angle of attack, an airplane climbs at its maximum rate only when there are no accelerations or decelerations in the air mass in which it is operating. In the situations in which these negative atmospheric phenomena eat into the pilot's margin against stall, his job is to supply a decrease in angle of attack or increase in power or both, *early on.* And above all, not to try to climb more steeply when the makings of climb just aren't there. Giving the airplane time and a better chance to accelerate is the way out.

HEAD-ON HORIZONTAL GUSTS

The Wright brothers are quoted as once having said that they would beat Glenn Curtiss in an upcoming contest because they knew more about air currents than he did. And they won—or at least Curtiss cracked up trying to fly on a day they considered too windy for flying. They were aerodynamicists; Curtiss was an engine man. Power is important, and he usually had a bit more of it than anyone else, but power doesn't take care of everything, particularly proper angle-of-attack control. With the limited performance of their Flyers, which at best flew close to the stall even in level flight, the Wrights sensed what we are only now coming to understand the full consequences of, namely, the effects of air currents in slow flight.

It's too bad their phrase "air currents" didn't survive. It seems to have gone down the drain with the newspaper reporters' odious phrase "air pockets." Air currents, though, quite accurately describes some of the elements of our modern term wind shear, such as head-on horizontal gusts.

After gradient-wind effects, head-on horizontal gusts are the most common undoer of good landings. That's what can happen when, just as we complete a round-out in our approach, we encounter a head-on gust, that is, one whose movement and acceleration are parallel to the runway.

When this happens, we balloon. If we sense the rise soon enough and decrease the angle of attack slightly, by lowering the nose a bit, we do not balloon very much. If we're not that quick with our countermove and the gust is prolonged, we may go from 10 feet right after round-out to 20 or 30 feet before we get the nose to a more nearly level attitude. That's when, in the new attitude, we'd better add a bit of power, because the gust gave us momentarily more airspeed and lift and drag, and the airplane has slowed down. When we fly out the backside of the gust, we will be slower than we should be at our new altitude. So, when a gust hits us, we quickly lower the nose a bit, gun it momentarily if the gust increases our altitude appreciably, and start our approach all over. Say from 30 feet above the runway.

The trouble with gusts after the flare in landing is that at an uncontrolled field we may not have been carrying any extra speed in our approach. Or at a controlled field, the gusts may be stronger than reported and consequently our approach-speed increase of half the gusts' velocity over the steady wind isn't enough.

In either case the result is the same because in the process of lifting us the gust also slows us down. With no extra speed in the approach, we come out the backside of the gust with less speed than we should have and can drop it in. Without *enough* extra speed, we can do the same thing.

If the tower reports the wind direction as along the runway at 20 with gusts to 30, we add 5 to our normal approach speed. By consensus that is enough to take care of the speed loss we will sustain during the encounter. But sometimes the gust at the runway threshold is stronger than where it is measured elsewhere on the airport. Consequently we fly out of the gust at less than normal speed after the round-out. Which might not be so bad if we didn't, on occasion, at this lowered gust-exit speed, fly right into a second and third gust, each of which further decays our airspeed and sets us up for a hard landing.

We have to learn to handle not one but a succession of gusts after the flare. In short, get in there and pitch. Yes, pitch and power. Less pitch the instant a gust starts lifting us, and if that doesn't do the job, power added in proportion to the higher perch we wind up on. We don't always have gust reports, and on windy days we must simply make a guess at how much extra speed to carry in our approach. It's not that the gust effect is so bad, but that we sometimes are late with countermeasures.

And drop in.

This term, downdrafts, has to be pretty broad. On final, without any input from the pilot, the rate of descent increases noticeably. What caused it?

It could result from flying into a mass of air that was settling, which would decrease the angle of attack momentarily and thus reduce lift, and we'd start down faster. Or it could result simply from a turned-down path in our headwind without any change in velocity, which would also change the direction of the relative wind and reduce angle of attack. Or, especially around small airports, which often have tree lines paralleling the runway and downdraft effects also from hills and ridges nearby, we can find ourselves on the express elevator going down.

Note, however, that in these situations the change in angle of attack has been a decrease, which is a plus as far as maintaining our margin below stalling angle of attack goes. So the stall potential has been decreased. But that still leaves us with an unwelcome increase in rate of descent.

Sometimes this triggers an impulse to pull the nose up some and at the same time increase the power. That would be all right quite close to the ground, say within 30 feet, but with more altitude, the main thing is not to try to outclimb the downdraft but to get some extra speed and get out of the phenomenon. Which means go to full power and even a bit less angle of attack or lower nose position to reduce drag and give the engine a better chance for acceleration as well as a chance to benefit from a forward component pull from gravity. Above all in this situation, a backfire in pilot thinking resulting in an attempt to outclimb the downdraft has to be avoided.

UPDRAFTS AND THERMALS

Updrafts and thermals are from below the horizon—they're below-the-belt relative wind effects—and they are more serious in approaches than downdrafts because the relative wind is coming from lower down and the angle between the relative wind and the wing is increased. This means an increase in lift, and even though there's no change in attitude of the aircraft, this translates into a flatter approach path. A feeling of ballooning is the first clue the pilot gets, followed by a slight loss in airspeed due to the increased drag caused by the new higher angle of attack.

In such an encounter, if the pilot is coming along on final with airspeed and rate of descent stabilized, flying at, say, a 7° angle of attack with a savings account of 7° above stall, the change in relative wind is going to make an immediate withdrawal of maybe only 2° or 3° from the account. The normal pilot reaction in this situation is to nose down a bit, which is correct.

But we must also think of the pilot who is coming in too slowly, with, say, an angle of attack much closer than 7° to stall. Enough of an updraft or thermal could undo him and produce a drop of the nose or lateral instability as the aircraft stalls. This happens only rarely simply because thermals and updrafts decrease in severity at the lower approach altitudes. But it can happen.

The most extreme example of this I know of involved a Brooks and Kelly Field graduate with a lot of Staggerwing Beechcraft time. He was flying a Cessna Airmaster at gross and, on downwind at a small airport and possibly a little on the slow side, flew across the burning slab-pile of a sawmill. The airplane dropped a wing and spun, in the snap of a finger. The accident made a deep impression on me because I had given this pilot his first flight in a Monocoupe and was much impressed by the precision with which he controlled airspeed and maneuvered the airplane, and, as rarely occurred on a first try, he landed it well. What the accident said to me was that superior training doesn't solve all problems and that a violent updraft at altitude is bad enough but can obviously be unmanageable at low altitude. Fortunately the principal result of low-level updrafts is simple overshoots.

UPDRAFTS AND ANGLE OF ATTACK

In considering updraft and thermal effects, even though they may not be a major troublemaker in approaches, we can stow away some worthwhile additional angle-of-attack information.

Since we do not normally have an angle-of-attack indicator, and instead use airspeed for angle-of-attack information, we need to think about the situations in which the seemingly logical correlation between angle of attack and airspeed breaks down. In normal operations, the rule high airspeed— low angle of attack/low airspeed—high angle of attack is valid except for the curves atmospheric effects can throw us.

Our airplanes are strong. But if, flying along at cruising speed, a pilot abruptly jerks the wheel all the way back, one or both wings come off. This is because the pilot had enough pitch control to put the wing at its maximum lift angle to the oncoming relative wind, and at cruise speed the lift produced exceeded the ultimate load factor of the wing. In every general-aviation airplane, there's a small placard somewhere in the cockpit naming the maneuvering speed of the airplane, which is normally quite a bit lower than cruising speed.

Flying along at maneuvering speed, which is the speed recommended for flying in turbulence, if a pilot pulls the wheel abruptly all the way back, he will almost go through the floor, but the wing at maximum angle of attack flying at maneuvering speed will not produce enough lift to give a higher load than the 3.9 G's the wing is designed to withstand in a normal-category airplane operation.

In the core of a thunderstorm, though, air can be moving vertically at up to 100 feet per second. A pilot flying into one of those shafts with cruising speed loses one or both wings as the G-load produced by the abrupt increase in angle of attack far exceeds the design load of the wing. If he hits the shaft flying at maneuvering speed, the airplane simply stalls, the nose pitches down, and the pilot's problem becomes trying to avoid exceeding the airplane's maneuvering speed in the stall and sometimes spin recovery and pullout.

If an airplane is dived to the redline shown on the airspeed indicator, hitting a bump that would normally produce only a moderate G-load increase would take the wings off. Certificated airplanes are dived to a bit more than the redline airspeed to see if control surface flutter can be induced by bumping the controls, but this is done in smooth air.

So, updrafts can get to us pretty easily when we're operating at the high end of our angle-of-attack range, but an instant cure is reducing angle of attack. Nevertheless, if they are severe enough, they can also get to us when we're flying at high speed and low angle of attack. Which stresses the importance of going to maneuvering speed when the bumps start trying to tell us something.

TURBULENCE

So far the principal vagaries of the dynamics of the atmosphere have been discussed in terms of changes in the angle between the relative wind and the mean chord line of the wing, and in terms of accelerations and decelerations of the relative wind. When it comes to turbulence, about all that can be said is that the air is badly scrambled and we can no longer think in terms of the probable direction of the relative wind or of any accelerations or decelerations in it, but only of the effect of turbulence on performance.

It is a wild and windy day and a little of everything may seem to be happening at once. Bumps. Lots of bumps, both up and down. The airplane has first one wing down and then the other. The roll-rate capability seems greatly diminished. The nose tends to hunt up and down more than normal. In some airplanes the tail does a lot of extra wiggling. And in the process in some airplanes, this may cause the wing tips to move in an elliptical path—with a combination of fore-and-aft as well as up-and-down movement. Sometimes this is described as a circular motion, but often it resembles more the outline of a lenticular cloud.

What a pilot thinks about at a time like this is anyone's guess. After all, how many textbooks are there on how to stay on a bucking horse? You just hang on and do what seems, and had better be, best from moment to moment.

The pilot in a turbulent-air climbout needs to concentrate on keeping the climb angle shallow, the wings level, and the heading constant. This, of course, can call for a lot of foot- and handwork, some of it coordinated and some not. The desideratum is whatever it takes in control movements to stabilize the attitude and heading as much as possible.

I asked my airline pilot/aviation author friend Bob Buck once (we were talking about turbulence in thunderstorm areas) whether he'd ever been at the point in big airplanes where he was worried about adequate controllability. His answer was no, but he added that he'd seen it bad enough that he felt if it got any worse he might be in trouble. In short, get in there and pitch.

So, in a turbulent-air climbout, we have to settle for less performance. How much less? There are not really any figures on this. Information about lift coefficient, stalling angle of attack, maximum lift angle of attack, drag, and so on come from wind-tunnel tests in which the big problem is getting a smooth flow of air over the model in the tunnel. Smooth air is necessary in order to get a usable unit of measurement. Otherwise it would not be possible to measure any improvement, for instance, in better streamlining or a more efficient wing section. Similarly, the performance figures in our aircraft owner's manuals are calculated from data points obtained in smooth air. Figures obtained in turbulence wouldn't mean much because one day's or area's turbulence would seldom resemble another's. We have to have standard units of measurement, in calm standard air.

One of the significant contributions to performance test flying was made by an early-day Boeing test pilot, Eddie Allen. The airlines had to have precise range figures for ocean flying. Allen discovered that on some flights the figures were too good to be true, and on others they made him wonder what was wrong with the airplane (or maybe the engineers).

What he figured out was that on some days he was flying in a vast area in which the air was rising slightly. On other days it might be settling; in other words, there was subsidence. On the up days he would cruise at a lower angle of attack and higher airspeed with normal cruise power. On the down days it took more power and a higher angle of attack to maintain altitude. He arranged for special meteorological observations in the area in which he was going to be testing and waited for days in which there was no lifting or subsidence to get his yardsticks.

To say that turbulence affects performance is only to say that it affects angle of attack. On a windy, gusty, rough day haven't you been climbing at your normal climbout speed only to have the stall horn give a toot? What it's saying is that you were closer to the stall than usual, which climbing more gradually at a higher airspeed would take care of.

TURBULENCE AND ANGLE OF ATTACK

For several years I had the good fortune to fly with one of Leonard Greene's angle-of-attack indicators (his company, Safe Flight Instrument Corporation, White Plains, New York, still makes them), which sensed movement of the stagnation point on the leading edge of the wing where, at different angles of attack, some of the air went over the top of the wing and some under. The panel unit had a left-right, slow-fast needle; when it was straight up the wing was flying at its maximum lift/drag ratio, or, in short, at its maximum-rate-of-climb angle of attack. On "bad" days, with the needle on the center mark in a climbout, the airspeed sometimes had to be 20 MPH higher than normal to keep the needle centered. And, of course, to maintain this speed, the nose of the airplane would have to be way below its usual climbout attitude. Which stressed that there is a good deal of elasticity in the correlation of airspeed with angle of attack in turbulent air. On a good day the Cessna 180 I was flying would show 800 FPM at 65 MPH at sea level. On a really rough day, the maximum rate of climb obtainable was sometimes as low as 500 FPM and it took 90 indicated to get that with the angle-of-attack needle straight up. Which said to me that on such a day, the airplane could easily stall if held to its normal climb attitude and airspeed, simply because the turbulence would make the angle of attack much higher than it would appear to be. For this reason during climbouts in turbulence we need to think the same way we do in approaches in turbulence, namely, in terms of an extra 5 to 10 MPH over the usual climb or approach speed.

And it follows, most surely, that we need also to think of the importance of some extra speed and lower angle of attack in turns. That is, I think, an even more serious necessity than in climbouts and approaches because we tend to operate closer to the stall in turns than in straight flight simply because we lack precise enough means of judging angle of attack in turning flight, our gauge being mainly only airspeed and/or G-load; the latter being a more certain indicator of angle of attack at low airspeed.

Yes, turbulence is an enemy, or at least it increases our vulnerability in slow flight and in maneuvering. In everything from C-3 Aeroncas and J-3 Cubs to Comanches, 210's, Bonanzas, and Twin Comanches time and again when I've circled an airport for a look at the wind sock and obstruction layout, something has told me, firmly, that in changing from upwind to downwind and crosswind, and in the bumps and rolls, it just wouldn't do to let the bank get steeper than 15°, and that I must keep the speed up. On these wild and windy days, gingerly getting onto the final approach course using shallow banks and keeping the speed up gets us into a straight-ahead-and-down flight regime in which the variables are more easily manageable, and a few extra miles in approach speed will enable us to keep our guard up.

TURBULENCE FIRST HAND

Turbulence affects also the altitude loss in a stall or spin recovery. Which is just another reason to operate much more on the conservative side on windy days.

I learned this fact of life on a wildly windy day at Evansville, Indiana. Flying a C-3 Aeronca, I had landed and started taxiing tail up. Fortunately, heading only slightly out of the wind took me into the lee of a large hangar and when I stopped moving the doors started opening most hospitably.

Inside, a fine-looking young man doing something to a new Waco F came over and after a circle of the C-3 said rather sneeringly, "Trouble with that piece of junk is that you can't fly it on a windy day."

His attitude wasn't unusual for an FBO in those days with a large investment in expensive training equipment. In his case he'd spent a lot of money getting a commercial at the prestigious Spartan School of Aeronautics at Tulsa and then, despite the recent national bank holiday, had been able to buy, at a high price, this elegant new Waco F intending to start a flight school at Evansville.

Well, I was an Aeronca factory representative (you could be a dealer if you could make the $400 down payment on the C-3's $1,412 price), and my answer was, "Well, I'll fly mine anytime you'll fly yours."

With that, he signaled the lineman and the hangar doors started opening.

The C-3 had no brakes and as I taxied downwind behind him I'm sure he felt I wouldn't make it to the starting line without weathercocking because he kept looking back. But the C-3 was quite manageable once you got on a downwind heading. If the nose started to the right, for instance, dropping the right aileron way down out there on that long wing would produce a yawing moment to the left on account of the wind pushing on the down aileron. At the same time, right rudder would expose quite an area to the following wind and that would also give a yaw to the left. A sort of reverse command operation.

The F took off in a very short distance and, there being

evidently a strong gradient effect, climbed almost straight up. With all my 36 horses in good fettle, I got off even shorter and climbed well enough.

At approximately 2000 feet above the airport the F leveled off and I waited to see what next. He did a loop, and then started down into an obvious pattern entry. So I thought a two-turn spin might be appropriate.

The C-3 Aeronca, while fragile looking, was a strong airplane, except for inverted flight loads on its cabane strut. On most days I'd spin it a couple of turns to show skeptics that it wouldn't come apart if the pilot sneezed. It never lost more than 500 feet in a two-turn spin and recovery.

But this time, when I closed the throttle and brought the nose up for a power-off stall, followed by full right rudder at the stall, it whipped as I'd never seen it do before into a fully developed spin. Normally this whip came only at the end of the first half-turn. I could tell something was up and immediately neutralized the controls. This had always given an instant end to autorotation. But today it didn't stop until the end of the first turn. I lost 1000 feet in the one turn and recovery, and in that descent learned a powerful lesson: the effect of turbulence on a stall recovery.

I made no sale. Even so, I was ahead for the day.

WHITHER BLOWS THE WIND?

It's possible that the preceding discussion of atmospheric effects might tend to alarm and detract from one's placidity. Which would be bad. So let's look a little at the bright side.

Atmospheric effects are not something which a pilot cannot cope with. They are simply factors that must be recognized and analyzed. Certainly the results cited do not happen every time. There are pilots who have flown for years and never have given any serious conscious consideration to the atmospheric effects on angle of attack and airspeed, or at least, assuming these things were obvious, have not talked about them. But our annual scrap heap of airplanes involved in takeoff and landing mishaps says otherwise. Atmospheric effects snare only pilots who fly too close to stalling angle of attack.

BIRD SENSE

Some pilots, from the start, react instinctively the way they should when they sense loss of lift or loss of airspeed. They seem to have what you might call bird sense when they encounter the relative wind coming up more steeply from below or faster: they instinctively nose more into the new relative wind, that is, reduce angle of attack. This is not just for the birds but for pilots, too, and it becomes the key move when the wing has suddenly been brought closer to stalling angle of attack without the pilot's having done anything to cause this. It can happen while climbing, turning, flying level slowly, or in an approach.

Rather than create an impression, however, that some sinister phenomena of movements within the atmosphere are just biding their time to throw a pilot, the intention is simply to make clear that a pilot can protect himself by thinking in terms of keeping angle of attack within its limited but safe bounds and without necessarily associating this with the attitude of the aircraft. As you have seen, in terms of pitch-change limits, the pilot doesn't have a lot to work with; and in turns especially, it means the waters get deep fast in banks of over 30°.

"I'll be wary" is a good motto. It means no steep turns when flying at a low airspeed unless the nose is kept well down, which means having enough altitude for a descending turn. And it means speeding up in downdrafts instead of trying to outclimb them. And it means not exposing you and your wing's bottom to a powerful gust in a steep turn at low altitude. And it means in takeoffs, approaches, and landings being sensitive to changes in climb rate, rate of descent, or airspeed which atmospheric quirks can produce. The key thing in dealing with these variables is making the necessary correction early on. Observe how often this means just a slight forward movement of the stick, or more power, or both. Whether you're going up, down, straight ahead, or turning.

I've always enjoyed flying on windy days. It means small airports become big ones what with the much lower ground speed on landing. And abetting the airplane's stability as needed is a not unpleasant link with the machine, for it makes the pilot feel needed. Windy days should be no problem if the obvious traps are kept at arm's length.

I think again of Bevo Howard. I doubt that he ever canceled one of his superb low-level demonstrations on account of wind. He may have allowed himself a bit more altitude or raised his minimum speeds in different maneuvers to allow for gusts and turbulence, but the show went on.

3

AIRCRAFT FLIGHT CHARACTERISTICS

There are some things about how different aircraft handle that help a pilot in takeoffs, approaches, and landings. There are also aircraft flight characteristics that do not make the pilot's job any easier.

Oddly enough, it often happens that some of the less cooperative aircraft are the ones most vociferously defended by their pilots. That may be because flying an airplane that requires something extra on the ball bestows status on the pilot. Or it may be that no matter how an airplane flies, when you get fully accustomed to it, you like it that way and tend to be defensive about it if the subject comes up. But the fact remains that nobody ever got on top of the sales heap building hot airplanes. The public is smarter than that.

An airplane can be made to fly just about any way the designer wants it to fly. And so I used to think that the industry was missing a sure bet in not picking out the one airplane that everyone seemed to fly best and finding out why this was true. Then they could make all their airplanes fly like that and thus expand the market. That one can go into the naiveté bin.

I think everyone in the design loop would concede that

while flight characteristics are important, the airplane is a complex machine and the production and the selling of it are even more complex. Not to mention the cost to develop and tool up for production of a new model.

AIRPLANE DESIGN: SCHMARTS AND LUCK

As well as the designer, the decision loop has to include production, marketing, sales, and in the end, their consensus reflects the conviction that it is not the place of the industry to tell the consumer what he should have, but to figure out what best meets the overall needs of any given market segment. If they don't manage to answer that need, the early and relentless answer is always the same: more money goes out than comes in. That can go on just so long. And so the arbiters of design are to be found not in the factories but in the marketplace. What you build either sells or doesn't, and it is the consumer who makes the decision. How "good" does the product have to be? Good enough to merit market acceptance. Too good and it doesn't sell, because it's too expensive. Not good enough and it doesn't sell, because it's not useful enough.

I'm reminded of a friend, Bill Kelley, who was at one time Sales Manager for King Radio Corporation. I always thought Bill had a talent for divining product salability.

He had a story about a company that was bringing out a new dog food. They assembled an array of experts to tell their top sales people how good the new product was. There'd never been anything like it before in nutritional value, health care virtues, vitamins, minerals, everything. It would make bad dogs good and good dogs better.

They had the crowd worked up as never before to go out

and sell, sell, sell. Wipe out the competition. Then a little fellow in the back row finally got the floor to say, "But the dogs don't like it."

The succeeding comments on characteristics of airplanes that help and hinder us in getting them gracefully off, up, around, and down are not intended as criticism of the manufacturers. They made them the way they are for what they felt were good reasons. And certainly in the case of the popular ones the reasons were good enough. Our purpose is simply to take them as they are and help the pilot figure out what he can do to fill the gap between the ideal and reality.

Even so, the ideal airplane is a pleasant thought. Some of the things that are not now available could become commonplace as a result of education of the consumer as to their desirability. But in the end the ideal airplane's cost and complexity (and maintenance) would limit its market.

There are many around who could contribute to the preliminary specifications for an ideal airplane. It should have a yaw damper; it should be spirally stable; it should have much better long-period longitudinal stability than is now customary; it should need no rudder P-factor or torque correction in climbout; it should have a very soft landing gear; it should have a three-axis autopilot with altitude hold and approach coupler; it should be fully deiced; it should be pressurized; it should have at least 1000-NM range at 75-percent power; if a twin, it should climb at least 500 FPM at 60 knots or less on one engine; if single engine, it should have two alternators and two vacuum pumps; it should have an encoding altimeter; radar;—are you still aboard?

Yes, those are some of the qualities an ideal airplane would surely have, but how about the pilots who want an airplane primarily to go 200 or 300 miles on the average, in good weather, carry one, sometimes two, and occasionally three passengers? In an airplane that is good on a short and rough field, has good glide-path control, and is reasonably easy to fly at not too high cost? Say with 145 HP, or a larger model with 230 HP?

In short, how about a 170/172 or 180/182 Cessna? That's the kind of consumer thinking Dwane Wallace, the moving spirit at Cessna, engaged in 30 years or so ago, and it put him on his way to meeting a $5 million payroll every Friday and capturing 60 percent of the general-aviation market.

Building and selling airplanes is a rugged business. Out of hundreds of inspired starters in aircraft manufacturing, frequently but not always undercapitalized, only three of the pioneers survived. Fortunately they did not all reach the same conclusions about what to build.

THE PIONEERS

Piper. Most price-conscious of the three. Build them strong but leave off the frills. They nourished the grass-roots market with the J-3 and moved up the size scale slowly, but always with the lowest price in any selected category, at some sacrifice in speed.

Cessna. Production and marketing were their primary strengths. They were first to recognize that as well as good tooling for quality control and lower production costs, there was no one-airplane solution for everyone, so a variety of models and horsepowers were vital. Keep them simple. Low maintenance. Good on run-of-the-mill airports.

Beech. Speed. They were the first to recognize that people would pay a lot for being fastest. With this performance goal, the airplane would be expensive in any event, so they added quality. Plush. The carriage trade. And they also zeroed in on the corporate market.

Had the big three started out all on the same track, we'd have missed not only today's rewarding development of the general-aviation market, but a lot of our best and most useful and most appealing airplanes would not be around. The big three would have fought over only a part of the aviation market, rather than offering choices that served to expand it.

To a considerable extent, the big three were never really competitive. And what a blessing. A Comanche for the fellow who couldn't afford a Bonanza. A Cherokee for the fellow more comfortable sitting atop a wing than suspended beneath it. A 172 or 182 for the fellow more comfortable suspended beneath a wing than sitting on top of it. And in fourth place, even if way behind the first three, Al Mooney sought speed at a price. Which Mooney Aircraft offers more today than ever before.

So much for the magnificently creative process that is every phase of general aviation and which has been alternately beneficent and cruel to those on whose lives it has impinged.

Let's turn now to our problems in getting good takeoffs and landings with our airplanes as they are. The better we know our airplanes the better we fly them.

HURDLES TO CLEAR

It seems that during the past 25 years, while general aviation has had its first real taste of dollar volume, our airplanes have gotten harder to land, or at least to land softly. Some may say this is mainly because performance has increased so much, but is it? Some of the high-performance airplanes are easier to land than some of the slow ones. It can also be said that good flight characteristics and ease of handling alone don't sell airplanes, else we'd not have had the boom we have.

Which suggests that pilots are willing to sacrifice some amount of their soft-landing ego for something they want more than that. Like more speed from minimum cabin cross-sectional area. But that doesn't eliminate a deep-seated desire in every pilot to make as good landings as possible in whatever he is flying.

OVER THE NOSE VISIBILITY

A high hurdle the pilot has to clear for good landings is the annual growth, small but steady, in the height of most of our instrument panels. This has come about because of the need for more panel space in which to put all the good things that make an airplane more useful: full gyro panel, dual navcom, transponder, DME, EGT, vacuum gauge, ADF, marker beacon receiver, and so on.

Once a fuselage jig and tooling are complete, you can't raise the roof a bit or widen the cabin for more panel space without spending a lot of money on new tooling and going through a new certification test besides. Which means a higher price, as well as less performance for a wider or taller airplane. So up goes the top of the panel, and away goes some of the windshield. Until in many of our airplanes, "Where's the runway?" is a good question in the final climactic seconds of a landing.

This lack of ability to see "where we're at" invites high-speed, near-level landings, with an occasional trip into the fence at the far end of the runway, or a too-late go-around, or an embarrassing drop-in. To be at our best and land at least a little tail-low, we have to acquire the ability to judge height by scanning around the left side of the nose, as best we can, and even at times by taking a quick glance out the other side during the hold-off. We can hunch forward and stretch our necks for maybe a helpful over-the-nose momentary peek at the runway ahead. Cushions and height-adjustable pilot seats help, but if these seats are adjustable enough to put our heads within an inch or less of the ceiling, in turbulence they can put a knot on our heads.

We have to see as best we can, and probably wouldn't want to give up 10 MPH cruise for a taller or wider cabin, which would permit an instrument-panel top that would give us a good view of the runway in a tail-low/touchdown attitude.

THOSE PEDALS

Before we get a shot at finding the ground in a skillful touchdown, though, we first have to get it off the ground, so I've gotten a little ahead of myself. Let's look at how to avoid knocking over one of the low hurdles in today's flying.

In other words, how much do we wiggle in the takeoff run?

Steering with foot pedals is awkward. If we had to steer our automobiles with our feet, the speed limit everywhere would probably have to be 20 MPH and the roads twice as wide as they are now. Yet with manual steering and drivers passing with only a foot or so between them at combined closing speeds of 100 MPH or more, head-on collisions and side-swipes are rare. And this is with drivers holding, by aeronautical standards, everything from Class I to Class X physicals.

Of course we wiggle, and sometimes it is a strenuous exercise to keep the nosewheel on the center line. For a number of reasons: a coarse control ratio is one thing; lack of feedback is another. As the power goes up, the tendency of the nose to swing left can be surprisingly strong in the initial stages of the takeoff run and its decreasing somewhat as speed is gained throws in another variable for the pilot to juggle. Some aircraft even have unstable landing gears. But mainly our steering problems are simply a question of the great difference between manual and foot dexterity.

Possibly some automotive engineer, skilled in the intricacies of caster and camber and whatnot, could give us a nosewheel steering system that would work as precisely as the manual steering of a car. But at a price, and with probably 25 to 30 pounds added at a very inconvenient place. The price might be so high, possibly a thou or so, that the supersteering would probably be offered as an option and the consumer would elect instead to continue with our crude but adequate systems of today. Let's just think of the situation as reflecting not a pilot's skill but simply the lack of a better steering system.

DIRECT LINKAGE

Some of our airplanes have direct linkage between the rudder pedals and the nosewheel steering arms, notably the Pipers. This has advantages. In soft ground or sand, it is at times possible to straighten up the nosewheel, where without direct control the wheel might get turned, start scuffing and bend something because full "rudder" wouldn't straighten it.

The disadvantage of a direct linkage to the nosewheel is that there's not a lot of rudder travel to work with so it takes a relatively strong push on a rudder pedal to initiate and continue a turn in taxiing. "Stiff rudder pedals" is a common term. The direct linkage also gets increasingly oversensitive if the takeoff run is continued to a higher speed than normal, as it might be purposely in a crosswind takeoff.

The direct system is, of course, an uncomplicated one and therefore less costly and it is also lighter. Maybe only a matter of a pound lighter, and so you may ask, why worry about the

---

weight? Let's have a glimpse of the engineer's world of ounces and pounds.

Taxiing a C-3 Aeronca once, I swung the tail over the tongue of a hay rake and caught a stabilizer brace wire enough to elongate the holes in the cadmium fittings on one side. Those fittings didn't weigh more than an ounce apiece. The steel replacements I had made at the nearest machine shop didn't weigh enough more to matter it seemed to me, and there'd be no elongated holes in those.

Back at the factory, I asked Jean Roché, the chief engineer and C-3 designer, why not stainless-steel fittings on the tail wires? The answer was a firm no. He explained that you have to be consistent about strength and weight, otherwise you wind up with an airplane senselessly overstrength in spots. Therefore it is overweight and will be a dog. Too heavy. No climb performance.

Maybe this philosophy explained why the C-3 would gain altitude in a series of consecutive loops, even with an inefficient wood prop and certainly never more than 36 HP. As William B. Stout, designer of the Ford Trimotor and Stout Sky Car, once said in a meeting, the most useful improvement that can be made in any airplane is to add some lightness. Roché didn't add it, he started with it.

SPRING STEERING

Rather than direct linkage, a more generally used system of nosewheel steering these days is one in which there is a spring between each rudder pedal and the nosewheel steering arm on its side. Push on a pedal and you're pushing on one end of a spring. As the spring compresses, it applies force to the steering arm to which it is attached. Which doesn't necessarily mean that the nosewheel has turned, but that it's being made to want to turn. With too light springs in the system, a pilot might say the pedals or steering is too spongy. Make them a bit stronger, and steering is much easier than with a direct linkage because any overcontrolling input doesn't take effect as quickly.

Pilots who get used to this type of control want no part of a stiff, direct linkage, and pilots used to direct linkage feel the spring system doesn't give them the positive control they want. The greater popularity of the spring system must be because pilots find they do a better job with it in keeping the line on the runway underneath the nosewheel on takeoff, and have less tendency to overcontrol as speed builds up.

FULL SWIVEL STEERING

A third system is a full-swivel nosewheel with no steering control whatever attached to it, as in the Grumman American Tiger and Cheetah. The steering is done with differential brake application. An advantage is that it permits pivoting on one wheel in a tight spot on the line. It also gives better control once the airplane gets up a little airspeed because the brakes are no longer needed and if not used the rudder gives, with feedback, a much more precise control—the kind one gets used to in taildraggers. The free swiveling nosewheel also eliminates the embarrassment of landing with the nosewheel cocked to one side as sometimes happens with a direct or spring-steering system.

A disadvantage of the full-swiveling nosewheel is that in a takeoff in a heavy crosswind, it may be necessary to run with one brake dragging to keep it straight until the rudder becomes effective, and that means it takes longer and farther to get to where the rudder becomes effective. With this system obviously brake wear is much higher than with the other systems. Also, the nosewheel has no shimmy damper as the other types do, and if the nut on top of the nosewheel swivel shaft is not tight enough the nosewheel can get to rotating or castoring round and round, which at least the onlookers find entertaining. But this doesn't happen if the correct friction is set into the shaft.

The engineers try for something better. One of the most unusual efforts was made in the Aero Commander Shrike series, with hydraulic nosewheel steering. Actually it works beautifully, but, for sure, not right away.

Airliners all have hydraulic nosewheel steering, because the strength in the captains' legs does not match the strength of their voices. The Shrike (or 500 series) is the smallest airplane so far to have it. With little or no information as to how the system works, however, a pilot is, at first, likely to go almost anywhere but where he wants to go. After a bit of taxiing around in the great open spaces, he's likely to say, "Oh, that's what you mean." Then he begins to like the system and use it with greater precision than he's accustomed to in nonhydraulic systems.

The valves controlling the servo cylinder's piston are connected to the toe-brake pedals on the rudder pedals. A slight pressure on one toe pedal, or in other words in a toe pedal's first movement forward, and the corresponding movement of the hydraulic piston is actuated. The trouble, at first, is that if you think, say, "Right brake, right turn" you're likely to move the rudder pedal forward in addition to applying pressure on the toe-brake pedal atop it. That means right rudder, right brake, *and* right hydraulic cylinder action, and it leads to overcontrolling.

The trick is to get the arch of each foot on its respective rudder pedal or bar and for a right turn, for instance, don't move the right foot forward but simply apply a light toe pressure to the brake part of the pedal while offsetting the tendency of the right pedal to move forward with a light pressure with the arch of the left foot.

This probably makes use of the system sound more complicated than it actually is, after a little practice. But the system gives better steering control in the takeoff run and less likelihood of overcontrolling. Like everything on the "good idea" list it comes at a price: dual hydraulic pumps plus an emergency nitrogen-bottle backup system, and there go the price tag and the weight.

LANDING GEAR STABILITY

Another aircraft characteristic that can make a subtle contribution to a pilot's secret frustration at not being able to steer his airplane as well as he steers his car is that our tricycle gears vary from positive to neutral to negative directional stability.

Taxiing along even at only moderate speed, if you give, say, the right rudder a good push and then put both feet on the floor, and the nose swings back even if only feebly toward the direction in which you were moving, you have positive directional stability in your landing gear. If, after such a procedure, the nose swings, then stops swinging but stays pointed in the new direction, the stability is neutral. If it keeps swinging, or tightens the turn, then you have directional instability. I don't know that this is a serious thing, but certainly if a gear isn't positively stable, it isn't helping the pilot as he races along the runway, ears back and a lot of everything happening all at once.

Oh, yes. The taildragger pilots: a minority group having it best of all. In the initial stages of their takeoff runs, their steerable tail wheel does a good job for them. Then as soon as they pick up the tail, it's all rudder and from there on they get even better feedback from an increasingly sensitive rudder than the tricycle pilot gets after lifting the nosewheel off. This is because in the tricycle, the rudder pedals must also overcome friction in the nosewheel steering mechanism, and in some cases the opposing forces of nosewheel centering springs or those of a fin on the tail of a nosewheel fairing.

Of course, the something-for-nothing rule still applies. The taildragger pilots, even with steerable tail wheels and brakes, can whip around in a ground loop after landing. It's too bad Dwane Wallace never wrote a how-to-fly book. It would have probably no more than 500 words in it, because he gets a phenomenal mileage per word. We once had quite a discussion on ground loops. Or at least I discussed and he summarized, thus: "You have to catch it early." That's all there is to say on the subject. No pilot ever ground-looped who got with it by the time the nose had swung no more than a few degrees. Got with it, that is, with opposite rudder and throttle, and more lately also opposite brake.

RIDING THE BRAKES

A final steering characteristic of our airplanes concerns the geometry of the brake pedals atop the rudder pedals, or the toe brakes. The extent to which they lean forward or back from vertical can be adjusted, and the best adjustment depends on the length of the pilot's foot and the height of the rudder-pedal bar above the floor. It also happens that in some cases the toe-pedal linkage is such that as a rudder pedal is moved forward, the toe-brake-pedal segment tends to move toward the pilot's toes. In airplanes that need a little help from the brakes in steering this is an asset and the pilot may taxi along thinking he is using only nosewheel steering when actually he is getting a combination of nosewheel steering and automatic differential brake application.

These things may not be so important to a pilot who is familiar with the airplane, but they can provide a degree of excitement when you put a pilot who isn't so familiar on the left side, especially if he has long feet. Even with his heels on the floor, which is always a good admonition for a visiting pilot, he is likely, unless he has his heels far enough back on the floor that his toes are off the toe-brake part of the pedals, to do his nosewheel steering in the takeoff run unaware that he is holding pressure on both brake pedals. If he gets into any degree of overcontrolling directionally, he's likely to apply even more pressure to both pedals as he walks the rudder.

This has happened twice to me in the recent past. In the first case the pilot was ex-military with a lot of jet time and the runway was long so I just sat it out and finally as the brakes got hot and soft it got off. In the other case, in a last-minute shift, I got a low-time pilot on the left side who, unknown to me, spoke only German. He started riding the brakes heavily early in the takeoff run, walking the pedals vigorously, and we just weren't accelerating properly. I'd have cut the throttle but he was leading a V-formation takeoff and I was afraid we'd get overrun in an abort. All I could think of was to keep tapping his knees and repeating *"Nein."* About the tenth time he tried to haul it off, it finally got off and that was that, but at our destination, by sign language, I opted for the landing.

This same sort of thing can happen with short-footed pilots if they happen to start their run with the arch of their foot on the rudder bar or pedal. In those cases, heels on the floor will take care of things and how far back from the pedals their heels are isn't critical.

Here's an example of how the consumer can do things the

engineers can't think of. In some of the Piper aircraft the rudder/brake-pedal assemblies are hinged on an overhead torque tube rather than beneath the floor. It took some foot lifting, but in one case a prospect got his feet up so high that he had the ball of each foot on the torque tube, and, pushing for all he was worth, went off the runway. I almost got caught on the same thing riding on the right side during a Seneca II demonstration flight when the pilot was slow to start a turn on a taxiway. Trying to furnish a little as-late-as-possible help with some left rudder, I joined the pilot momentarily and also pushed on the torque tube. Until then I'd thought that the fellow I'd heard about who pushed on the torque tube instead of the brakes was stupid.

This is no way to think in general aviation, though, until you've been there. Once out at Grumman's on Long Island, one of the production test pilots landed gear up. One of the Gillies brothers, Bud, I think, active in test flying and a major stockholder, called the production test pilots in and gave them a forceful lecture on there being no excuse for a gear-up landing and warned that such carelessness would not be tolerated in the future. He then went out and got into his private F-3-F and flew over to the Aviation Country Club for lunch and landed gear up. There were subsequent gear-up landings at Grumman, but nobody got fired.

SPRING, SPRANG, SPRUNG

"Hope springs eternal," or something like that.

Some airplanes have an interconnecting spring between the aileron and rudder controls. Bank a little to the right, say, and during the roll period while the ailerons are displaced, right one up, left one down, the spring moves the rudder slightly to the right, unless the pilot holds the rudder pedals centered, which he isn't supposed to do in a roll. Rolling into and out of turns, this spring interconnect gives good rudder coordination and thus eliminates any need for control coordination by the pilot. In this respect it is a good thing.

The spring coordination of rudder and aileron displacements is not, however, the reason the spring is in the airplane. For a look at why it is there, let's start with the regulation under which a certificated airplane must level the wings out of a turn if the pilot uses rudder only. Some airplanes won't quite cut it because of their lateral/directional couple—insufficient yaw capability for the amount of dihedral in the wing, with ailerons centered and rudder full over. Changing the rudder stop to permit greater rudder travel might take care of picking up a wing with rudder only, but maximum rudder travel is a touchy thing. Too much and flat spin problems develop.

The reason for the regulation on unbanking with rudder alone, as I heard it, was that an inspector of the Aeronautics Branch of the Department of Commerce took off for a test flight (the Branch at that time did the final certification test flying) in an airplane in which the aileron cables had been disconnected. Fortunately the airplane had a rudder/dihedral ratio that permitted him to bank and unbank and maintain wings-level flight with rudder alone, so he got it off, up, around, and down without incident though not without anxiety. He was able to sell the Branch on the idea that all airplanes should have this capability. Obviously it is worth having if one ever needs it.

There is also a back side to the coin, which can be handled in the same manner as the front side, provided one can perceive the problem properly and control himself.

An Aeronautical Corporation of America compatriot of mine, James C. Welsch, later became a Cessna dealer for a time on Roosevelt Field on Long Island. One day he rented one of his Cessna 140's to a military pilot. Certainly a physically perfect and well-educated pilot with a year's ground school and $200,000 worth of flight training didn't need a check-out in a Cessna 140. But as the pilot took off, friend Welsch had sudden and painful misgivings about that conclusion. Just as the airplane became airborne the right wing dropped, the wing tip contacted the runway, and the airplane cartwheeled. Fortunately the pilot was unscathed, but not the airplane. Trouble was, the aileron cables had been crossed during a just-completed inspection.

After some months the airplane had been rejuvenated and Jim decided to test-fly it. In running the checklist he moved the wheel from side to side and observed that the ailerons moved, but as he said later, not necessarily in the right direction. On takeoff he also went into a roll, dug in a wing tip, and rolled it into a ball a second time. The aileron cables had again been connected crosswise.

So it came to be, for a while, that a certificated airplane could be banked and unbanked with rudder alone. So if the ailerons didn't work at all, or if they worked in reverse with normal control movements, and you could remember to turn loose the wheel quickly enough, which isn't likely, the airplane could be unbanked with rudder alone.

REDUNDANCY?

That, at least, was the original concept. But in time it eroded somewhat. Some airplanes wouldn't quite unbank with rudder alone, and approval was given to use of a rudder-/aileron interconnect spring that would feed in a bit of appropriate aileron travel when the rudder was moved.

So far, so good, or almost so. But, alas. With an aileron/-rudder interconnect spring, if you keep the wings level in a climbout and hold enough right rudder to keep the ball in the middle, or take off with enough right rudder trim to do the job, the spring is going to put pressure on the ailerons for a roll to the right and it becomes necessary to hold left-wheel pressure to keep the wings level. This gives the feel and seemingly the geometry of flying with crossed controls: right rudder pressure or right rudder trim/left aileron pressure, but actually it isn't the real thing. This is not too bad in a single-engine airplane because the control coordination the spring gives in the rest of the flight is worth the price. Just don't let feeling a bit awkward in the climbout disturb you.

With an interconnect spring it is likely that you'll come to practice a pilot's common concept of the conservation of energy: instead of holding right rudder and left aileron in wings-level climbouts, you'll wind up with the right wing down a bit and no right rudder pressure. This saves energy because holding a rudder displaced requires a steady muscle drain, but putting a wing down just far enough to hold the runway heading calls for no muscle input once the degree of bank is achieved.

There's nothing in flying that pilots don't finally figure out how to do with the least effort. And in this case there may be a bit of lagniappe thrown in for a right-wing-low climbout. I don't know that there's ever been a scientifically acceptable test to determine whether an airplane with the engine canted slightly to the right will outclimb one with the engine pointing straight ahead, but Fred Weick canted the Ercoupe engine to the right because he had no individual rudder control, and he was able to climb out wings level. I've always wondered if for this reason the Ercoupe would dust anything comparable with the same horsepower.

SPRING TEST

The way to check for a rudder/aileron spring is to make a taxiing turn with hands off the wheel and see if the wheel moves to "bank" the airplane in the direction of the turn. In some tricycles the spring is strong enough to make it possible to steer along the taxiway with wheel only. With the airplane standing still, moving the wheel full over and releasing it to see if it moves back toward neutral may or may not indicate a spring interconnection with the rudder. For instance, the Piper Warrior, which does not have a rudder/aileron interconnect spring, has an aileron-centering spring system, presumably for increased feedback, which returns the control to neutral.

In summary, a rudder/aileron interconnect spring in a single-engine airplane that has noticeable adverse yaw from aileron displacement does a lot of footwork for a pilot. In the climbout it may make him wonder if gremlins are responsible for his having to hold the right wing up while holding right rudder in the climb to keep the ball in the middle. Or he may fly the line of least resistance and climb out right wing low.

Redundancy is, of course, a form of life preserver in an airplane and thus is always desirable, but the rudder/aileron spring provides redundancy only if it can move the ailerons. Truly redundant lateral control is available only if an interconnect spring isn't needed. Interestingly enough, had he needed it, an interconnect spring would not have saved the Department of Commerce Adam who conceived it. Nor his issue.

If you're lucky you don't need brains, but if for either reason you become wealthy and are convinced that flying with two engines is twice as safe as flying with only one and your dream airplane has a rudder/aileron interconnect spring, it can give you a fit flying on one engine unless you have three-axis trim. Few general-aviation light twins (and no singles) have this feature.

The onus of a rudder/aileron interconnect spring in a twin comes in two packages. The first comes when you lose an engine in the climbout. The directional trim situation is conventionally taken care of by a heavy foot on the rudder on the side of the good engine as necessary to hold the runway heading, and then using rudder trim to get the load off your leg. With the spring, the resulting rudder deflection causes the ailerons to move and roll the airplane toward the good engine unless the pilot counters with wheel movement in the opposite direction, or, to be more precise, wheel pressure as necessary to keep the ailerons centered. In thus "holding the wing up," the rudder pressure necessary to prevent yaw is increased, via the spring.

In an earlier day one of the precepts of engine-out operation in a twin was to carry the live engine a few degrees down. This reduces the rudder pressure, or rudder deflection, necessary to prevent yaw, which is a help, but with the spring, no matter whether you hold wings level in the climb or carry one wing low on the good-engine side, you wind up with some trim toward the live engine. This means flying with rudder displaced, which puts a spring load on the aileron to roll toward the live engine. You have consequently to override the spring by holding the wing "up" in order to keep it from going farther down, and this can be a chore considering all the other pressures and cockpit duties following loss of an engine in the climbout. In this situation, lateral trim capability is a blessing, if there's a spring.

The second part of the spring onus comes with an en-route engine failure or when you're getting into level flight after an engine-out climb. If you fly wings level, you will need some rudder trim in the direction of the live engine, and that means having to hold a wing up all the time. Without a spring it would be possible to trim toward the live engine until you were carrying the live engine low and you'd fly straight without roll. But with the spring, in attempting to trim with rudder only to a wing-low attitude, you wind up with some rudder deflection and this means that even in wing-low cruising flight you have to hold a wing up with ailerons. While the spring plagues the single-engine pilot only during the climb, the twin pilot has an ague in both climb and cruise. And it can cause unbelievably quick changes in heading, if the pilot's on instruments, changing frequencies, looking for an approach plate, or rechecking the altimeter and an MDA or DH. Seems to me in a twin a spring is a poor fix.

CONTROL-WHEEL TRAVEL

This modern aircraft flight characteristic you'll have to give the ladies some credit for. While a control stick is more readily understood and gives less friction in the aileron control linkup, it had disadvantages. Or at least it did when skirts rather than slacks were fashionable.

Moving the stick forward and up underneath the instrument panel provided a happy solution to this problem. A sprocket and chain were put on top of the stick and the shaft of the control wheel was run through a bushing in a socket in the instrument panel and attached by hinge to the aileron control sprocket.

Sometimes the man as well as the woman pays. As the stick up underneath the panel moves fore and aft, its tip describes an arc and this moves the attached end of the wheel control shaft up and down. Consequently, the wheel moves up and down between full forward and full back.

The key element is that in some aircraft, in its last inch of back-travel, the wheel moves up quite a bit as well as back. If the shaft of the wheel is considerably higher than the pilot's elbows, this upward movement can cause the pilot to pull down as well as back on the wheel in landing, thus greatly increasing the load on the bushing and consequently the friction on the shaft. If the bushing happens to be dry, or of a nonroller type, the pilot may, in trying for a tail-low touchdown, pull until the force required is so high that he thinks he has hit the up-elevator stop, although he hasn't. Thus he can fail to give it that last bit of up-elevator which so often puts the feathers into a touchdown.

If, as a hostage to fashions of yesteryear, there is any possibility that you're failing to have the wheel really all the way back on touchdown, run a cockpit check on the ground. Pull the wheel full back a few times and observe how much it rises. Then on your next landing lift up a little on the wheel as you go for a stick-back touchdown. This can make a surprising and most gratifying difference in your landing if your control-wheel shaft comes out of the panel on the high side, and especially if you're involved with a strong downspring in the elevator-control system.

BREAKAWAY FRICTION

Another hindrance to becoming a wheeler-dealer in landings is breakaway friction in the elevator-control system. In cruising flight or in a climbout, there may be enough vibration to cover up most of this characteristic. But in an approach, with low power or power off, control friction can cause jerky use of the elevators.

You flare and shortly thereafter want to increase the angle of attack slightly. You try a little back-pressure on the wheel and nothing happens. So you add a bit more back-pressure and all of a sudden you go through the breakaway poundage required to get the wheel moving and wind up with more back-wheel movement than you wanted; and so you zoom a bit.

Somehow this breakaway-friction phenomenon seems to peak in the trainers, or at least most of the ones I rent. On the ground sometime you might want to attach a fish scale to the control wheel and see how many pounds it takes to make the elevator start moving. If the figure seems high, try getting the elevator hinges oiled, the control cables checked for excessive tightness, and any pulleys and hinges in the system lubricated. Also, the control-wheel shaft bushing in the panel may need attention. The lower the control friction and breakaway friction the less porpoising you'll experience while waiting for a touchdown.

Ball-bearing control systems would, of course, make a big difference, but you can imagine what that would cost today. So let's just have a nostalgic sigh for the old Fairchild 24, which had them.

If you're having a control-friction handicap that an OPEC deal won't cure, and you don't have a too heavy downspring to deal with, you may find that, on occasion, pumping the wheel—making very short and high-frequency movements of the wheel fore and aft in the hold-off—will give you the feel for lift you're alerted for but aren't experiencing. This is a heresy in officialdom and there was a time when pumping the stick this way would fail you on a flight test. The verdict was that such activity revealed a lack of proper feel for the airplane. In government by bureaucracy, rules get on the books with the greatest of ease and are almost impossible to remove. Since this one was made, things have changed a lot. Downsprings, more friction-prone control systems, less feel available. If a small bit of judicious stick-pumping helps on occasion, there's no reason not to explore that channel in order to get better acquainted with your airplane. It can also tell you that you're flying too fast after the flare and can anticipate a longer-than-usual float, or that you're too slow and may need a touch of power.

THREE-AXIS TRIM

Someone once said that one of the main lessons of fairy tales is to be mindful of what you wish for: you might get it. Nevertheless, I've long wanted to own an airplane with controllable aileron and rudder trim in addition to the standard elevator trim.

When airplanes were slower, seating was tandem, and the gas tank was in the fuselage behind the engine, careful rigging and use of adjustable (on the ground) trim tabs on the ailerons and rudder, as needed, gave us at least some airplanes that would fly straight and level until they hit a bump and a wing dropped.

Things are different now. We have side-by-side seating, fuel is in the wings—at times also in tiptanks—the speed range is greater, and we have micromass production. Production test pilots with 10 airplanes a day to fly have to check a lot of things in a very short time. Come windy and bumpy days, they're hard pressed to do a good job on checking the rigging and trim. They may come down from a flight and use a trim-tab bender with the handle on it for a slight change in the angle of the aileron or rudder tab and walk away. It's surprising that under such conditions they so often come mighty close to getting it right.

I'm convinced, however, that many an aircraft owner, including yours truly, has flown a slightly wing-heavy airplane for many a day without giving it much thought. In rough air it is not very noticeable. You're busy anyhow picking up first one wing and then the other. It's only in smooth air that you can really gauge the situation by simply turning loose the controls and waiting to see what happens.

Today this condition can be aggravated by more fuel in one wing tank than the other, or a heavier-than-thou passenger on the other side. Or, conversely, such weight imbalance can correct for an airplane that is wing-heavy flown solo. Finally, though you think time and again en route that after you land you're going to give that left aileron tab a bit of up-bend and the other one a touch of down, somehow the mental tape erases on touchdown and you don't get the job done. Which may be just as well.

I had an airplane once that I flew for a year, making this resolve every time I had a smooth-air flight. When I finally did give the aileron tabs a slight change in angle to pick up the low wing it was a bit too much and the airplane changed from right to left "wing heaviness." A touch of left bend in the rudder tab to correct this? That done, it again flew right wing heavy. For which I overcorrected with some more left aileron trim. So back and forth. Too much or too little of one and not enough of the other, and vice versa. A lot of "trim" flights.

Finally, my goal became to get back to the start, if possible, and settle for carrying a light left pressure on the stick to hold the right wing up, or else a light left rudder pressure. Either of which, of course, is tiresome on a long flight.

This frustrating exercise, which is likely to hit almost any-one in a first venture in trim-tab bending, may be a primary reason why manufacturers prefer to avoid having more than an elevator-trim control in the cockpit, if that is feasible. If not, then just one extra trim control, for the rudder. There is also the consideration of the extra cost of rudder- and aileron-trim controls, either with trim tabs or a bungee trim system. But mainly I believe the consensus is the fewer trim controls the better, because pilots who fly the smaller airplanes would probably have more trouble with them than without. Also, it probably did not go unnoticed in the trade that a bungee aileron-trim system available by an aftermarket STC for Cessna's classic Skylane did not sell.

In short, granted a three-axis trim wish, you might find that getting an airplane accurately trimmed was not the simple thing it appeared to be. And I'm sure when it came to the occasional pilot, which most of us are, and rental airplane, many an FBO would find his airplanes back on the line with the ailerons trimmed full over one way and the rudder trimmed the other way. Which might give straight and wings-level flight but at a speed loss due to the deflected control surfaces. Or, the FBO might worry about some timid soul taking off with full aileron trim only and digging a wing tip.

HOW TO SET YOUR THREE-AXIS TRIM

At Santa Monica once, where I was seeking more information on Bill Lear's new Super ARCON wing leveler, I met Ed Conklin, general manager of the LearCal Division of Lear, Incorporated.

In a discussion about autopilot/trim problems Ed passed on to me a nugget from his extensive experience, in the Air Transport Command in World War II, watching hundreds of pilots use three-axis trim systems. Invariably, when their airplane wanted to turn, either from being wing-heavy or out of trim, they'd start by centering the turn needle with rudder trim and then get the ball in the middle with aileron trim. This was a sort of echo of the prehistoric 1-2-3 system of instrument flying: 1) Stop the turn (with rudder); 2) center the ball (with ailerons); and 3) check the airspeed (with the elevator).

Actually, trimming this way can finally get the job done but it can take a long time. Ed said he'd seen pilots go back and forth between rudder trim and aileron trim for 30 minutes before they got the airplane willing to at least try to fly straight and wings-level.

Ed had an entirely different system of trimming an airplane. Be sure not to miss the crucial point that in his system the procedure is just the opposite of the foregoing more instinctive system.

The Conklin three-axis trim system: 1) Turn the airplane loose; 2) trim the ball to the center with rudder trim; 3) trim the turn needle to center with aileron trim.

This is an almost instant, precise method of three-axis trimming. When an airplane wants to turn and you let it start to do so, it seems logical to pick up the low wing with aileron trim, but the fact of turning doesn't tell you anything at all about the possible rudder-trim situation. So, the first chore is to get the ball centered with correcting rudder trim, which eliminates any tendency for yaw and rolling moments. From there on it's an easy and positive step to center the turn needle with aileron trim. So, *ball* centered with *rudder* trim, *needle* centered with *aileron* trim. It's as simple as that, and in a matter of seconds your airplane is trimmed.

THE CASE FOR THREE-AXIS TRIM

Except for the heavier twins, we are still in the wishing stage on three-axis trim. The pilots of the heavier twins don't even have to ask for it; anybody building big airplanes knows they're going to be flown a lot on instruments, when lack of three-axis trim capability can be extremely burdensome. During a moment of frequency changing, or a glance at the altimeter setting or the MDA or DH figure on the approach plate, the heading can change 20° or 30° in an out-of-trim airplane before you know it. Today not only the small twins but many single-engine airplanes are flown on instruments regularly.

Many of those pilots do not realize how much less work they'd have to do if they had three-axis trim. Seems it should at least be offered as an option.

What about VFR, which is most of the time for most of us? I think it is needed then, too. Flying an airplane in which it is necessary to hold a bit of aileron pressure to keep out of a turn, or, if one prefers, a slight pressure on one rudder pedal, is not only a lot of work, but it can make a pilot feel there's something sinister about an airplane that has to be flown this way. He may even become tense and frustrated if he doesn't understand it's not that the devil is after him but simply that he's flying an out-of-trim airplane. After all, one of the most difficult achievements in flying is a seemingly effortless ability to fly straight and level. That ability is difficult to acquire in an out-of-trim airplane. And if a pilot is ground down to a nub en route by holding a wing up or pressure on a rudder pedal all the way, he's not likely to be able to provide the degree of concentration and perceptivity needed to make a good approach and landing.

But we're getting there, slowly. While our trainers and the most popular four-place airplanes have only elevator trim, and pilots are obviously happy enough with them, the rest also have at least controllable rudder-trim systems. Piper sort of led on this. They do the best they can to get each airplane rigged as well as possible, and in flight you simply crank in rudder trim as necessary to make the airplane want to hold a heading. That means winding up flying one wing low in almost every case. The speed loss is probably negligible unless the airplane is quite wing heavy, and it is a simple system that leaves most pilots contented.

Of course when you fly this way, the little airplane in the artificial horizon is going to be flying one-wing-low, and the IFR pilot has to remember that's "level" or at least is the lateral attitude it will take to keep out of a turn. If he forgets and holds the wings level, then he's going to go into a skidded turn unless he maintains pressure on one rudder pedal or retrims the rudder. And if he does that, he's going to have to remember to hold the heavy wing up constantly with ailerons to keep it level. It's less work to use rudder trim to hold a heading and remember which wing to fly low because that doesn't require a constant pressure on the aileron control.

AUTOPILOTS VERSUS TRIM

Present-day autopilots for small general-aviation aircraft have contributed to lack of consumer demand for three-axis trim. They tend to sweep the need under the rug. Their development has had an interesting history.

The first autopilots, primarily for large airplanes, were three-control; that is, they controlled rudder, ailerons, and elevators. They were pioneered by Sperry, Bendix, and Lear, and they were expensive and heavy.

By the early fifties when general-aviation pilots began to think more in terms of the transportation value of their airplanes, there were, fortunately, entrepreneurs around who sought an acceptable balance between economics and basic autopilot needs. These people understood the weather problem in cross-country flying and anticipated correctly that many a VFR pilot, especially flying the new and much faster machines, was going to have spiral-dive problems. Their worthy goal was a low-cost wing leveler of under 10 pounds—an aerial life preserver without precedent, and even today a minimum for single-pilot IFR operation.

The wing levelers were first advertised as safety devices, and there were some pretty lurid spiral-dive depictions in the advertisements. This left most pilots cold, though, because no one thought he would ever get into a spiral dive. But the failure of the safety appeal wasn't the reason the wing levelers failed to find a market. The problem was that the primary buying motive was not safety but desire for something to make flying less work. The wing levelers didn't quite do that, or at least not well enough.

Nevertheless, there was a bumper crop of wing levelers: Dave Blanton's Javelin; Bill Lear's ARCON; the Globe from Dayton; Don Mitchell's Airboy; the Brittain, financed originally by Dr. Karl Frudenfeld, were the leaders. The basic concept was to actuate a rudder servo with a high-speed rate gyro so that a yaw signal would produce an opposite rudder movement until the yaw stopped, or, more precisely, until it moved the rudder enough to keep a turn from really getting started. That meant the wing would be kept level or else would be picked up in a turn or spiral until there was no longer a yaw signal, which would mean therefore that the wings must be level.

In all cases, however, these good people stumbled over lack of three-axis trim. Pilots found as they flew along they would have to keep using the trim switch to keep on a heading satisfactorily.

And so, after some highly competitive years in the single-axis, or single-control, autopilot field, the first question of each wing-leveler prospect began to echo. Invariably they asked, "Will it hold a heading?"

ROUND TWO—HEADING LOCK

Before long the manufacturers all arranged their systems so that an overriding signal from a pickup on the directional gyro would tell the otherwise rate-gyro-controlled servo that picking up a wing wasn't enough. It would have to turn the airplane as necessary to get back on heading. Pilots really went for that feature, and the phrase "heading lock" was born.

But in this flying business more is never enough, and today our relatively low-cost and lightweight two-control (ailerons and elevator) autopilots have not only heading lock but pitch control and altitude hold (from a pickup on the altimeter), and will take us from omni to omni, to and along the localizer, and down the glide path!

With only rudder or aileron trim, but not both, to give them a cooperative airplane to fly, they are fine tuned to keep the handicap of not having three-axis trim to a minimum. Virtually all of the airplanes these autopilots are installed in have rudder trim now; but with only that, or in the case of the Bonanza only lateral trim, they have to fly the airplane wing-low unless the manual trim is set precisely before the autopilot is turned on. It's no sweat to the user of the autopilot to fly this way, but the autopilot sweats in doing so, and some models even have an automatic cutoff feature if a lateral imbalance is putting too great a load on the aileron servo.

The bright side is that the users are happy and therefore unconcerned about the intricacies of three-axis trim, for which their autopilots have removed them from the potential market. It is indeed a blessing to be able to trim an airplane as best one can and turn on one of these two-control autopilots and contemplate the beauties and wonders of flight rather than the mechanics of it. They provide, incidentally, another example of the consumer being unwilling to pay for something that doesn't do what he wants it to do. Which makes finding out what the consumer wants, and can pay for, the secret of success in our free-enterprise system.

The case I'm trying to plead is that of the pilot who does not have the luxury of a two-control autopilot to enjoy. If he had three-axis trim, he would find himself much less a hauler of water and hewer of stone and would get a refreshing taste of the joy of flying an airplane in perfect trim.

The increasingly common use of downsprings in elevator control systems is not making today's airplanes any easier to land. A downspring can be conveniently visualized as a coil spring one end of which is attached to the instrument panel just ahead of the elevator-control wheel and the other end attached to the wheel itself. The farther back the wheel is pulled, the more the spring is stretched and the harder it is to pull the wheel back. Downsprings, in fact, are usually near the tail cone, attached to either a down-elevator cable or elevator-control arm and to a fuselage member.

Downsprings are used for two purposes: to increase the approved rearward CG limit; and to increase the stick-force gradient and thus, it is hoped, to reduce the likelihood of a pilot's pulling back too hard on the control wheel coming out of a dive and overloading the wing structure, as he might more easily do in an airplane with light elevator-control forces.

The use of a downspring to increase the permissible rearward CG limit permits a bigger load in the rear baggage compartment, which is a highly desirable feature. But in doing this the spring, presumably, increases the longitudinal stability. Which confuses me. You see, when flying slower than trim speed, by regulation an airplane has to nose down with the wheel turned loose, in which situation the elevator would be in trail. These mandatory nose-down characteristics are normally achieved by aerodynamic means. In short, static longitudinal stability in pure form is an inherent characteristic of a certificated airplane, and a pilot does nothing to add to or subtract from it.

Unless he overloads the rear baggage compartment enough, or even, in some cases, just the rear seats. In one classic case not so long ago a pilot was overoptimistic about his estimated rearward CG situation on a cargo flight. He rotated at his normal rotation speed and right after leaving the runway the nose started gradually rising on its own, even with the wheel full forward. Instead of having any longitudinal stability, the pilot had a divergence, or instability. So he cut the power and kaboom landed flat and climbed out of the washed-out airplane.

Almost every manufacturer has lost an airplane, and sometimes a pilot, in the process of finding out just how much rearward CG can be specified in the owner's manual. They keep loading the airplane tail-heavier and tail-heavier. All of a sudden it's too much, and right after spin entry the use of full-forward elevator-control-wheel movement doesn't pitch the nose down and break the stall. The spin chute on the tail is tried, and if that doesn't pitch the nose down farther the pilot bails out. Then, on a second prototype, they go back to the farthest rearward loading that permitted proper stall/spin recovery and, after rechecking this, specify a maximum rearward CG an inch, at least, farther forward than that in order to protect the pilot who has trouble with multiplication and addition in the onerous mathematical calculation of center of gravity. And also to provide a margin for the pilot who thinks the, say, 200-pounds-maximum-weight placard in the baggage compartment means he can carry that much on every flight, and doesn't realize that that weight has nothing to do with permissible rearward CG but is a structural limit on what can be put in the compartment *if* the rearward CG is not compromised by doing so.

How does a downspring contribute to permitting a more rearward CG? It seems to me that a downspring gives an appearance of increased longitudinal stability without actually adding any. The test pilot determines a rearward CG that wouldn't be quite acceptable without a downspring. There's a little but not enough longitudinal stability left at the stall with the wheel free. But add a downspring and turn the wheel loose and the spring moves the wheel forward enough to get a pitch down. Somehow this seems to be crossing the line between the aerodynamics of inherent stability and the mechanics of control. At any rate, downsprings are approved and legal and their use widely accepted in practice.

Once a downspring is strong enough to accomplish its rearward CG increment feat, it is sometimes made still stronger to increase the stick-force gradient. In order to be certificated, an airplane has to meet prescribed minimum standards of stick-force gradient. Stick-force gradient means simply that as you move the wheel back gradually to fly more slowly than any given trim speed, the amount of back pressure required must increase steadily the farther back you pull the wheel. If the amount of back-pressure increase, or stick-force gradient, is small throughout the wheel back-travel range, pilots say the airplane has light elevators. If the increase is considerable, they say the elevators are heavy. Too light elevators and it is easy to overcontrol in pitch, especially in turbulence or in a high-speed pullout due to the anxiety that comes into play when the ground starts coming up fast. Too heavy elevators —thus a seemingly headstrong airplane—and concern about proper controllability may be a normal reaction, at least on first acquaintance with an airplane. But there's controllability enough, or the airplane wouldn't be certificated. It is more important to appraise what extra-strong downsprings deprive the pilot of in approaches and landings.

In approaches, stick-force gradient can have an angle-of-attack indicator function. Trimmed for a bit more than the approach speed he wants to use, the pilot learns that a certain light back-pressure gives the speed he wants, which frees him to look at what's ahead and to monitor his approach flight path, instead of spending too many seconds looking at the airspeed indicator. But with an extra strong downspring, the stick force needed to fly even a little more slowly than trim speed is going to be high and the higher it is the harder it is to judge and hold what's proper, due to muscle fatigue. Consequently the pilot trims for a little lower speed to get the load off the wheel and thus is deprived of the early signal that lighter elevators would give him of changes in airspeed if he came along on final able to carry only a moderate back-pressure on the wheel to maintain his approach speed.

Someone once said about warning devices that you have to be looking to see a light but you don't have to be listening to hear a horn. (He'd never landed gear up.) That's nice phrasing and is especially applicable to what a not-too-steep stick-force gradient says to a pilot about angle of attack, for he has neither to be looking nor listening to get the message. It can come to him hands-on right through his arms in every approach and landing. When he gets to know his airplane after a few hours in it, he develops his own private muscular/memory monitor of approach speed, given a not-too-steep stick-force gradient.

In this situation he is getting feedback as to elevator response and consequently an angle-of-attack indication as well as change in angle of attack. Feedback, or lack of it, is what you get when you start to open a door by pushing on it. Suddenly the knob practically goes out of your hand because someone on the other side is also opening the door by pulling on it. This startles you because, instead of the expected feedback from hinge breakaway friction and the inertia of the door, you're experiencing instability. What you were expecting was first a push to get the door started moving, then an increase or decrease in the push as needed to make the door move at the rate you want. The feedback is in the effect on you of changes in the amount of push required to open the door at the rate you want to open it. If it doesn't take much and you overshoot a bit, you reduce the push. Not enough and you push a bit more. If an airplane had a negative stick-force gradient, it would be unstable and inherently uncontrollable.

The reason landings with extra-strong downsprings are more difficult to do well is that as you round out, carrying approach trim, and then go for a wheel-back tail-low touchdown at the moment you run out of altitude, you may find it necessary to be applying as much as 15 pounds of back pressure on the wheel in some of our airplanes. Having to pull that hard tends to desensitize you to the feedback from the stick-force gradient, and most especially when you're obeying the FAA dictum of always landing with the right hand on the throttle.

On slow airplanes, stick-force gradient is controlled by the length of the elevator-control arm. As top speeds go up, gradient control is achieved not only by mechanical arm length or leverage ratios between the control wheel and elevator but also by the aerodynamic balance of the elevators. Which means that part of the elevator area is ahead of the elevator-hinge line so that in an up-elevator position the balance area will dig down into the airstream and thus lighten the load on the control wheel. On still faster airplanes, lead is sometimes added to the tip of the balance area to reduce the possibility of elevator flutter at high speed. Finally, today, the downspring is added both for rearward CG increments and to provide a restraint on the pilot if he should jerk the wheel back in some high-speed situation. The downspring, of course, is a strictly mechanical restraint based solely on wheel-back position and is not related to airspeed or any other aerodynamic input. It is at its strongest with full up-elevator at touchdown. Which brings us to the question of how to land happily ever afterward with downsprings.

Each airplane that uses downsprings is going to be different, but a good example of what a pilot needs in order to get a downspring-handle on, and while we're at it, some other characteristics in many of today's high-performance airplanes, came out rather clearly in a recent flight in a Mooney 231. It cost me $300 to fly the airplane for one hour. I rented the plane at $75 an hour from the nearest Mooney dealer, who happened to be 250 miles away. That's three hours traveling time for one hour local flight, and it works out to five dollars a minute for my time in the air!

The airplane was a beauty, with just about King Radio's entire option list, autopilot with coupler, built-in oxygen system, turbocharger, deluxe interior, and, importantly, a $900 electric trim switch on the control wheel. The 231's beauty is more than skin deep. It cruises at more than three times its stall speed, thus giving us more and faster miles per gallon on considerably less horsepower than anything else of competitive speed. And when you look beyond some of its rather special flight characteristics, it has soul.

PREFLIGHTING DOWNSPRING STRENGTH

It is well to get an early introduction to the strength of the downspring in any airplane. This is easily done during a cockpit check. On the 231, centering the trim indicator pointer in the takeoff band on the trim scale and then pulling back on the control wheel took quite a lot of force. So there's the downspring. This is the pull necessary to rotate in the takeoff run, plus whatever the airload on the tilted-up elevator adds to the control force required.

Next, set the trim needle about halfway up the scale toward "Nose Up," to estimate where it might be on final: it takes still more pull to get the wheel back an inch or so. This is the force that would be necessary to flare with approach trim, plus, again, whatever the airload on the up-elevator adds.

Finally, pulling the wheel all the way back with approach trim required quite a lot of force, which is what could be expected in control force just before touchdown, plus airload.

It doesn't bear solely on the touchdown phase, but as a matter of interest, set the trim all the way nose-up and then push the wheel all the way forward. The airplane also has what might be called an upspring, which you can feel even though it is comparatively weak. Mooney designer Roy Lopresti is a miser when it comes to drag. If a designer wants a bit more longitudinal stability, the usual process is to increase the area of the horizontal stabilizer. I was once told that if you have a speed-associated stability problem, just keep increasing the size of the tail and finally you'll get the airplane slowed down enough that the problem goes away. But that's obviously not for Lopresti. He left the stabilizer area alone and stabilized the elevator with opposing springs, which, in effect, adds effective area to the stabilizer. All for a few more of those hard-earned extra miles per hour with no increase in power.

In my first takeoff, with the trim set just above the center of the takeoff band, and 10° flaps as recommended, with a passenger front and rear, it took a pretty good pull to get the nose up a bit at 75 to 80 MPH. Shortly the airplane became rather gracefully airborne. The recommended initial climb speed for forward CG is 90 and for rearward CG 95. We reached 90 a second or two after leaving the runway, and I decided to try to hold that speed, but the back pressure required was so high I pulled back on the electric trim switch. Slowly this got the load off the wheel. Raising the gear produced no appreciable change in attitude. Raising the flaps for the airplane's lively rate of climb produced a slight sag and also called for a bit of nose-down trim. Now, the thing is, there's aerodynamically no funny business going on with this nose-heaviness in the climb. Without the downspring, the airplane would probably not be noticeably nose heavy, but the elevator responsiveness might be a bit too much on the lively side. The nose-heaviness, then, is not an airplane but a downspring characteristic.

TRIMMING FOR TAKEOFF

It was tempting to experiment with a more nose-up trim setting for takeoff, but that takeoff mark on the trim scale certainly reflects the best judgment of somebody in the know about this airplane. It seems likely that the recommended takeoff trim setting is for around 110 MPH, which is the best rate of climb speed, flaps up. If the trim were set more nose-up for takeoff and a pilot got a lift-off and minimal nose-heaviness in the initial stages of the climb, he might easily go into a too steep climb, especially if he overtrimmed nose-up. A pilot can be counted on to pull a lot more than he can be counted on to push on the wheel. With the recommended takeoff trim set, the airplane is telling the pilot he's way below the best rate of climb speed, and it keeps reminding him to speed up. It seems best to use the takeoff trim recommended and pull on the wheel as necessary to hold a moderately nose-up attitude until we're well out of the ground cushion and have the gear and flaps up. Then trimming at 90 in the climb would be good practice, and shortly thereafter 110 if the climb is to be continued. Having trimmed for 110 in the climb, we'll find it necessary to push and trim quite a bit in the level-off as the speed builds up.

DISCOVERING AN AIRPLANE'S INDIVIDUALITY

Some airplanes pitch down when full flaps are applied, some pitch up, and some do not pitch appreciably either way. It all depends on the change in the downwash pattern of the wing in the stabilizer area. Putting the gear down caused little change in the 231's trim. Even putting the flaps down 10° didn't disturb the attitude appreciably. But from there on, flapwise, another special feature of the airplane surfaced more emphatically.

When full flaps are applied, the 231 pitches down, really down, which in itself is not too bad. But in arresting the strong and rather abrupt pitch-down, you trip over the downspring and have to pull hard and start trimming. The flap switch is spring loaded and has to be held down to keep the flaps going down. If hand trimming should be necessary, it might be best to alternate between flap switch and trim wheel. With the electric trim system, holding the trim switch back with the left thumb while holding the flap switch down with a right-hand finger gives the servo motors a good chance to wind up in a no-pitch-change tie. But it seems best to start this after turning on final, gear down, 10° flaps, and trimmed for 90 MPH, since the right hand is going to be tied up—or rather down —with the flap switch.

I found that in a simulated balked landing with gear and flaps down and trimmed for 90 MPH, adding full power caused just a pleasant nose-up trend.

Finally, to give those springs in the elevator-control system their due, when you're just flying along, small corrections in pitch aren't often needed to supplement the short-period disturbance corrections the airplane makes automatically, and when required, only a light momentary wheel pressure is needed. The springs are unobtrusive en route, and that's most of one's flying time.

By now I had experienced two of the 231's three most special flight characteristics. The first is the role of its downspring on stick-force gradient: i.e., it takes a good push or pull to fly faster or slower than trim speed. Which is mainly the springs talking rather than the airplane. The second special quality is the strong pitch-down tendency as the flaps are lowered past ten 10°. The answer again is learning to lead with the trim tab. Now it was time to experience the last of the triad: the way the Mooney floats, and floats, and floats out of a proper approach-speed flare.

The first landing was probably about standard. With partial power and 500 FPM sink rate and trimmed for 90 MPH, and full flaps early, since they come down rather slowly, it was steady enough as she goes.

Then the flare and throttle closed. It took a heavy back-pressure just to accomplish the flare and then seemingly even more to hold the attitude while waiting for sink to start. I couldn't keep off the trim switch. And it floated and floated. Even though I thought I had gotten full nose-up trim during the float (I hadn't, due to a seat belt–trim wheel tangle), it was still very nose heavy at the touchdown, tail low. The touch-down occurred about 1500 feet beyond the approach end of the runway, which had been crossed with about 20 feet elevation.

CORRECTING FOR A LONG FLOAT

Anytime an airplane floats excessively in a landing, it is logical to think that it will be easy to eliminate the extra float simply by coming in more slowly. But in the case of the 231, this is not the answer. It floats not because you're too fast but because of its short landing gear, which results in the wing being so close to the runway that in the hold-off the air packs up underneath the wing and causes a pressure rise and an increase in the density of the air under the wing. The seagulls know about this ground-cushion phenomenon. Barely skim-ming the surface of the water, they go a hundred yards or more seemingly without moving a feather.

Rather than trying to eliminate or reduce floating with a lower approach speed, it's better to stick with the book speeds for approaches; otherwise you'll be squandering the 1.3 times flaps-down stall speed, which protects against the changes in angle of attack that can be produced by the vagaries of the ambient air in the approach and over the runway.

In short-legged low-wing airplanes that have floating proclivities nearing the bottom of the ground cushion, it is important to recognize that the characteristic doesn't even start until the wing is within the last few feet above the runway. So that's the elevation at which an answer has to start working: how to get it on the runway sooner so the brakes can be used.

The most popular procedure, which helps a little but not really enough, is a fly-it-on type of landing. This means coming in with a bit more than prescribed approach speed, flaring quite close to the runway into a slightly nose-up attitude, and holding that attitude and letting it land itself. Usually, the airplane settles on, not too hard but firmly, in its near-level attitude. This can get a touchdown somewhat less far along the runway than is possible with a standard-height flare and hold-off for a tail-low touchdown, but the near-level touchdown is at a much higher speed and consequently the turn off the runway is at about the same place, but with a lot of extra wear and tear on brakes and tires.

A second and somewhat more effective procedure, which reduces both floating and stopping distances, is to go into the ground cushion with a higher sink rate than normal, that is, by coming in with prescribed approach speed but with sufficiently less power to give an extra 100 FPM rate of descent. This calls for a faster and more nose-up and thus more drag-producing flare deep in the ground cushion and gives an earlier touchdown at a lower speed and consequently shorter stop than in the first procedure. But this still isn't the shortest landing possible out of a proper approach speed.

A third procedure, and my preference even though it is

controversial in some quarters, is to eliminate the extra float by raising the flaps gradually after the round-out. But first, right after the round-out into level flight is finished, it is important to wait a second to see if the airplane is really going into a float so as to allow for any transient atmospheric effects.

In my second landing in the 231, intending this time to raise the flaps after assurance that it was starting to float, I couldn't find the flap switch and with so much back pressure needed to hold the nose up with approach trim I didn't dare look. I hadn't forwarned him, so when I called out, "Get the flaps up," I was surprised that the 231's guardian, Derwood Martin of Raleigh-Durham Aviation, obliged. As the flaps came up slowly, it became progressively easier in an acceptable time sequence to get the nose up in the hold-off, and the touchdown came with the wheel all the way back, and reasonably soft. Right after touchdown, with the wheel held all the way back the nose bobbed up gently about 5 degrees and then settled back, wheel still back.

With only moderate braking, it was possible to make the first turnoff, which is 1800 feet from the approach end of the runway, slightly up-grade. The approach end of the runway had been crossed with just under 90 MPH and about 20 feet elevation. Density altitude was right at 2500 feet. Incidentally, this turnoff is the best I can do with a Cutlass or 152-2 under similar light-wind conditions, except for a little lighter braking.

In the next landing I was able to find the flap switch without looking and right after round-out started both the trim servo and flap servo running, giving me full nose-up trim and flaps down at about the same pace. With the lighter control force and better feedback, the touchdown was softer than in the previous attempt.

Two more similar landings were about the same. Same

turnoff with only moderate braking and a bit softer touch-down each time. I also found that in these wheel-full-back touchdowns, if the wheel was moved after the touchdown just a small bit forward, the nose would not bob up.

Finally, jackpot. Power-off approaches have gone out of style, but I do not like to fly a single-engine airplane, or even a twin for that matter, without finding out something about its radius of action power-off.

LAND POWER-OFF AND LEARN

In a rather pontifical tone the tower gave their blessing to a power-off 180° side approach. With the gear down and trimmed for 100 MPH right opposite the approach end of the runway on downwind, I cut the power and asked Derwood to stay with me.

It wasn't bad at all. The by now accustomed flap-and-trim sequence and coordination to avoid a tug of war with the wheel worked perfectly. Ten-degree flaps and load trimmed off the wheel at the start of the 180° turn. Then trim and half flaps at the 90° point. Airspeed still 100. It looked good. But on reaching the runway heading and lineup, flaps-down and in trim, it seemed we were a bit low and while the extra 10 MPH over normal approach speed might have made it possi-ble (Derwood thought so) to stretch it to the runway in a trade-off of speed for distance, a short touch of power seemed best since there is sometimes a downdraft at this end of the runway due to terrain slope. Then, round-out, start of the float, trim switch all the way back and flap switch held all the way up and wheel gradually all the way back. I'm not sure I ever want to fly a Mooney again because it might mean spoil-ing the memory of such a blissful landing. I'd never landed any softer. We coasted slowly past the 1800 foot turnoff and taxied to the parking area.

The 231 flies just like any other airplane, granted an anticipatory understanding of its downspring, flaps-pitch-down and float characteristics.

FLAPS AND FLOAT

Raising the flaps to reduce the float is a controversial procedure. It is easy to see why. The first objection is usually the possibility of raising the gear instead of the flaps. As far as the 231 is concerned, though, the gear light is for the first time where it ought to be: near the top of the instrument panel right in front of the pilot. The flap switch is at about four thirty and way down low. So the probability of landing gear-up from this procedure seems minimal.

Nevertheless, an owner should go at the procedure gingerly, remembering that first of all the flaps are left alone unless and until there's obviously going to be a floating problem. Then he can start by experimenting: raising the flaps only a little at first, and next time a bit more, getting the nose a bit higher each time to offset the loss in lift. The touchdown is going to be at flaps-up landing speed finally but it is sooner and the landing ends up much shorter.

Now to the practice of trimming the load off the wheel where there is a heavy downspring load on it and touching down with full nose-up trim. The principal objection is what the trim situation would be in a last-minute go-around. To discover the muscle required for this operation, it is only necessary to get into a power-off glide at altitude, gear and flaps down, with full nose-up trim, and bring in full power. You need to be sure that the forward pressure needed on the wheel could be handled with the left arm only in a go-around, since the other hand could be busy with throttle, prop, trim wheel, or whatever. Rather than trimming in the round-out and hold-off, the opponents say it's better to just go ahead and land with approach trim and learn to pull as necessary. Which is all right, but I doubt that it will give the percentage of soft landings available with nose-up trim where the down-spring is extra strong.

If a low-wing is short legged enough, it tends to float. My first experience with this was in the Piper Comanche, then, to a somewhat lesser degree, the Twin Comanche. It is also a characteristic of the Seneca II, especially if even a trace of power is left on after the flare. I also floated in a Citation I and the earlier Learjet; or at least during my single landings in each of them, they seemed to float a bit much after using the prescribed approach speed. Which made me itch for a touch of up-flap or speed brake or something to cut the float. But at Citation and Learjet prices one does not dare experiment!

I also know that in the characteristically heavily flapped Cessna high-wings, unless you come in much too fast they don't float very much. But in any event don't fool with raising those high-lift flaps: even if it does float, in doing so it can set you down hard and forthwith.

WHERE CREDIT IS DUE

The flaps-up caper I call the Strohmeier system, after William D. Strohmeier, one of the brightest lights in general-aviation marketing and long-time Piperman. Bill's flaps-up sequence in making short landings in the Comanches was, you might say, born of necessity. They floated out of normal-speed approach round-outs. The flaps were manually controlled with a long lever, with a ratchet button on top, that you lifted from the floor to lower the flaps. Hearing that Bill could flare and immediately make a soft tail-low touchdown without a float, I decided to try to catch up. It was easy. If the flap lever was let down too fast, it could be raised slightly and held there to correct the sink-rate situation. In no time I had it. Flap lever slowly down and wheel slowly back, and you were on tail-low with only an acceptable amount of float.

When it came to the Twin Comanche, which had an electric flap switch, not spring loaded, I didn't try the flap trick for quite a while, afraid that the rate at which the flaps came up might be too fast and I might be too busy to stop them by centering the flap switch. In time I found the rate was just about right, so it was just a case of moving the flap switch to the up position and then trimming nose-up with the electric trim switch on the wheel to get some of the load off the control wheel. In some years of flying Comanches and Twin Comanches, I never got a drop-in from raising the flaps, never did a go-around, nor did I ever turn off closer than 200 feet to the end of a paved 1957-foot, 6-inch runway I frequented. Which is only to say that the Strohmeier system is easy and has merit for short-legged low-wing airplanes. Bill is now domestic adman for Aérospatiale's pacesetting Airbus, but from where he sits I don't think he can get to the flap switch.

STIFF-LEGGED GEAR

A final hurdle on the course to soft landings, and a high one, relates to the trend toward stiff gears, mainly in the retractables. Time being a factor in shock absorption, a soft gear tends to be a long gear. Tucking a long gear up is often impossible due to structurally limited space problems, especially in airplanes that started out with fixed gears. Also, a long-legged gear means more weight, more power needed to retract, extra gear-door complications, more cost, and usually less gear-down performance. So, short and stiffer pays dividends, except for the soft-landing buff.

The ultimate in short, and not really too stiff, is the Mooney gear. It's a compact little marvel, as it has to be considering the thin wing root into which it has to tuck. The Bonanzas have a soft gear. A good example of the difference between long and short gears and oleos is in the Cherokee Six 300 and the Lance. Soft landings in the Six with its long-travel oleo gear are a cinch. The same airplane with a retractable gear and less oleo travel—the Lance—requires a lot more finesse to get a soft touchdown. The Aero Commander 112 or 114 with its trailing-beam oleo gear set a new high in soft-gear retractables. And so on.

What can a pilot do about a stiff gear? Well, not a lot, but at least he shouldn't regard the frequency of firm to hard touchdowns as a personal shortcoming. My only suggestion is that with a stiff gear it pays to learn to complete the round-out closer to the runway than is normal. That way, if it drops in, it hasn't as far to go and the bounce won't be as high. It appears there are some retractables in the works with soft and forgiving gears, which will be a welcome advance.

4

PILOT CHARACTERISTICS

Having considered what the dynamics of the atmosphere can do to angle of attack and our attempts to fly with precision, and how certain aircraft characteristics make our efforts more demanding, it's now time to explore how the pilot flies: what he does to the airplane, what the airplane does to him, and what he sometimes does to himself.

OUR PHYSICAL ATTRIBUTES

It has been said that from being groundbound these many eons, human beings have not developed the senses they need to operate well in the air. I do not think this is true. Certainly we all do not have what it takes to play par golf, win at Wimbledon, receive gold medals in the Olympics, or wind up in the Baseball Hall of Fame. But by the millions we play golf, tennis, and all the other sports and enjoy them to the fullest. The same five senses that serve so well in these endeavors are more than adequate to supply our physical needs in flying. Let's have a glance at them.

In our takeoffs we learn that a certain nose-up attitude gives us our best lift-off. Maximum rate of climb? We soon learn this is related to a certain pitch attitude, granted power available, to add a bureaucratic hedge. Pretty soon in each climb-out, we learn to hit it about right and then fly on standby for perceiving any pitch change that might be caused by turbulence, or by failure to hold the nose up or trim it there.

We get these messages of pitch and pitch changes not necessarily while staring over the nose. It is possible to look straight down at a flat piece of ground and still judge pitch attitude and any change in it. Even while we're looking to one side for traffic in a climb, the wing's angle to the horizon tells us if all is well or changing. Even in a turn we can get needed attitude monitoring information from the angle the wing makes with anything vertical in our line of sight along the bottom or top of the wing—a tree, the corner of a building, a flagpole, a smokestack. And, of course, our eye tells us bank angle in our turns. At first we can't, but somewhere close to a private certificate we begin to see drift clearly when close to the ground. Later we learn to see it at altitude, too, simply by getting on a selected compass heading and lining up two points on the ground directly ahead and then doing whatever it takes headingwise to keep those points in line.

We look at spume on water, the sway of wind patterns in wheat fields, the movement of leaves in treetops, and the movement of smoke and cloud shadows on the ground to tell us about wind direction and velocity. Which is important because impact forces in a downwind landing vary as the square of the ground speed.

On final, incredible! Our eye tells us whether we're too high, too low, overshooting, undershooting. Imagine the computations as to angles, changes in angles, change in rate of change in angles, and perspective changes of the runway that are silently racing through the brain and that finally give us a motor sensory message for a little less, or more, power, a bit more nose-up or -down, or a notice simply to hold everything. The gods never had it any better.

Finally, as we flare, in our peripheral vision we know how many feet down the runway is and whether we're settling toward it too rapidly or too slowly, or at just the right rate. And thus how and when to change back pressure on the stick, and at what rate.

FEEL

Deep muscle sense, sense of touch, or whatever you prefer. This one is a wonder of wonders, too. We sense from the feedback in change in stick forces whether we're moving a control too slowly or too fast to get what we want. This sense also includes, somewhere inside us, a combination accelerometer and G-meter. Some call this seat-of-the-pants flying, but I think the sensor is more visceral than that. Maybe the "black box" is in the vicinity of the solar plexus.

We sense accelerations, fore and aft, as in the takeoff run, when manifold pressure and a constant-speed prop can conceal lack of normal thrust. We can also sense lateral accelerations, as in a skid or slip. Sensing a vertical acceleration tells us we're in an updraft or downdraft. Excessive G-load in a turn can tell us we're tempting the fates.

In an approach we sense any change in rate of sink, and this tells us to add power and how much to use. In approaching and in a stall, we sense sinking and loss of lift, and though we don't always do it, the way out is both a reduction of angle of attack via forward stick movement and an increase in power if it is not already at maximum. This signal is priceless because it is more immediately vital to safety than any other, and we need be neither looking nor listening to get it. This sense of accelerations is essentially our monitor of *changes* in angle of attack, and that is where we often get caught short.

HEARING

They used to talk about listening to the angels sing: the sound of the flying wires vibrating on the old biplanes when the approach speed was too slow. (Cessna does it today with a hole in the leading edge of the wing that allows the "wind" to blow on a reed.) Let's not kid ourselves. The wind noises are largely gone as well as the wires, and good riddance, but we still hear more than we think that is important and helpful. An extra whistle on takeoff may tell us a door is not properly locked and not to panic if it opens and sounds off with a loud and steady roar. It can also be a reminder that this distraction often results in wheels-up landings in our haste to get back on the airport.

Many a pilot has made a precautionary landing in time simply from having detected an unusual sound in the engine exhaust, or a new whine in the gear train in the accessories section; or, by combining sound and feel, has detected a vibration change that could involve the engine, propeller, or the airframe (like an open gear door, or open oil-filler access door, or inspection plate loose or off).

Come, now, you might say. But yes. An oil leak that drips on a cylinder head or exhaust manifold lets you know in a hurry. The acrid odor of electrical insulation cooking is unmistakable. An overworked gear motor can emit both sounds and smells. In an 8:00-A.M.-to-2:00-A.M.-next-morning flight from Roosevelt Field to Wichita I was unaware that my Culver Cadet did not have a voltage regulator. Fifteen hours into the flight I did notice an odd acidic odor. As I found some weeks later, the battery had boiled over and the acid had eaten through the battery box and eliminated a lot of termites who'd been holding hands inside one of the longerons.

So, the word is sniff if there's any new scent.

TASTE

This one I'm not so sure about, except that it might be a backup for our sense of smell. Some things can be smelled and tasted simultaneously, as I learned once 1300 feet down in a salt mine. Or, with a cold, one can taste more surely than smell. Encountering my first dust storm in western Texas, I was not sure what was going on. Accustomed to the New York area's combination of fog and smoke, I first thought it was something along that line. But pretty soon, though I couldn't smell due to a cold, I could taste the good earth, like slightly muddy water out of a tap.

Thus, our five senses. If they sound a bit primitive, consider the services they perform for a juggler, a ballet dancer, a gymnast, or even a lowly handball player. Or how, in our aeromedical dark ages, they enabled us to stand on one foot, eyes closed, for half a minute, in order to be considered fit to fly.

OUR SIXTH SENSE

Sometimes mention is made of someone seemingly having a sixth sense. Usually this is called intuition. Igor Sikorsky's *Story of the Winged S* has a priceless chapter entitled "Imaginative Intuition." He tells how in his early airplane-building experiences, circa 1910, he would pick up a wooden airframe part, raise and lower it and turn it, and decide that it didn't feel quite strong enough. As later became evident, he was right. His intuition had permitted him to reach into the unknown, beyond the state of the art in mathematical stress analysis. In retrospect he reasoned that these fortuitous analyses happened so many times, it simply could not have been by coincidence that they were so often right.

If intuition is a sixth sense, it is more mental than physical and I do not know what part it might play in getting our airplanes up, around, and down safely and gracefully. But it does seem that if there is a sixth sense that belongs in place beside the five mentioned earlier, it is a sense of time. We may not be able to say with much accuracy when five minutes or ten minutes is up, or even an hour, but in shorter lengths, we may be sensitive to the fourth dimension.

We flare for a landing, hold our attitude waiting for the settling to start, and it doesn't start. Flying level at this juncture we're bound to be slowing down with power off, or no change in power. So why don't we settle?

It could be from a momentary increase in wind along the runway of just the right velocity to offset the loss of lift that our otherwise decreasing airspeed would have produced. Obviously this can go on just so long, not to mention that the wind velocity can drop back to what it was when we completed our flare.

Our sense of time tells us that the normal time sequence from flare to settling has, for some reason, been stretched and that when settling does start, it is going to be faster than normal, requiring a faster-than-normal increase in back stick-pressure and rate of change in nose-up attitude. If our sense of time isn't up to par in such a situation, we drop it in.

OUR MENTAL ATTRIBUTES

The most basic thing about pilot flight characteristics is not lack of physical attributes, but rather a lack of understanding fully situations in which, from a combination of fear and instincts, we do wrong things with the controls.

FEAR

This one is a bit touchy. No pilot likes to admit to being afraid of flying, or of an airplane. That would sound cowardly. But a smidgen of fear is, I think, a necessary ingredient in a safe pilot. Without it you have the overconfident pilot, who always seems to wind up in trouble.

But our fears, whatever they are, must be controlled. The main problem is that in flying we run afoul of a basic fear, the fear of height, the fear of falling. It is a powerful fear, maybe even more overpowering than the fear of asphyxiation, or of loss of sight, or of fire. Maybe this is because we are more in contact with it on a day-to-day basis, in walking, jogging, going down steps, and all the other activities that can give us broken hips and broken arms when we try to break a fall. It is the basic fear a child first learns from climbing too high. It never leaves us.

As we tell people who say they don't want to fly because they have an acute fear of height, they won't experience this paralyzing sensation in an airplane because the input is optical and the ingredients aren't there. There's nothing coming up from the ground to the airplane that tells us, visually, that we're atop a flagpole.

This is a fortunate physical fact for pilots as well because it provides the first step in the repression of our fear of height. We get used to being up there and it doesn't bother us. Or at least most of us; I've flown with pilots who'd been flying quite a long time, yet put on the left side in a new airplane, or one that had a reputation of being hot, and their knees would shake, their heels dance on the floor, and sometimes their hands even shook. What these signs meant to me was that here was a pilot who'd not adequately repressed the fear of height. As a result flying was too stressful to allow him to operate at the full level of mental activity and perceptiveness of which he was otherwise capable.

We tell students who are about to squeeze the knobs off the control wheel to relax. Telling them to relax when they're scared is confusing cause and effect. They're tense because their muscles are straining, and the cause of that is in the head. What is needed is some rationalizing about flying risks. Statistics can be helpful. Not the overall statistics, but the ones that apply to a reasonably prudent person's flying.

Fear of engine failure is a high hurdle for all of us. Engine failures are a minor contributor to accident causes. In single-engine aircraft there's a three-out-of-four chance of no more than hull damage in a forced landing, and the fourth chance characteristically involves a pilot-induced spin-in.

Pilots who squirt the tanks and gascolators before the first flight each day and after each refueling rarely have an engine failure on takeoff.

Many forced landings result from pilots trying to keep going when the gauges have long been flirting with the "E" mark. They don't get into trouble from making a precautionary landing because of low fuel but from trying to stay up with no fuel at all.

Engine failures due to structural failures in the engine are the last thing of all to worry about. There are a lot of comforting figures around.

THOSE REASSURING NUMBERS

If it's not engines, it can be something else that's gnawing on a pilot. I found my relief in statistics. There weren't many figures available in the late twenties, but the official record was a fatal accident every 10,000 hours in civil aviation. Structural failures and fire in the air appeared to be primary causes. I was able to find out that most of the structural failures occurred when pilots who'd never had any training in aerobatics tried to do some. As to fires in the air, fatigue failures at the flare of the copper tubing in its attachment to the carburetors was a common cause, plus, in some cases, a red-hot exhaust pipe running right underneath the carburetor. Flexible fuel lines were becoming available and appeared to be solving the broken-fuel-line problem.

Finally, what about the frequent newspaper reports about some famous pilot getting knocked off? If they couldn't make it, what chance did Joe Blow have? It didn't take a great deal of local inquiry into those cases to see that these pilots were stretching their luck. They'd become prominent or famous by pushing the state of the art to its limits, or beyond. Racing, aerobatics, distance flights, overloaded takeoffs on short fields, and such. They often sacrificed security for publicity, not always solely for their benefit but in a belief and common desire to contribute something to the advancement of aviation.

When I decided that the safety problem involved primarily not the airplane but the pilot, I started flying.

My feeble search to get the stress level in everyday flying down to a comfortable point translates into a much clearer picture today. There is more information available now, and most of the accidents you look into simply make no sense at all.

Someone taking off at 2:00 A.M. on a foggy night with little night time and no instrument capability and going in only a few miles from the point of takeoff isn't an accident. It is a suicide. And so are the cases of the buzzers who in the pullout from a dive collide with something they don't see, or who, in a spectacularly steep zoom, usually in a low-performance and sometimes overloaded aircraft, wind up with the nose snapping down in a whip stall and dive into the ground because they hadn't room to regain enough speed to pull out. And so on.

If it will help, people who don't fly half or more potted, people who don't buzz, and people who don't fly cross-country at night or in poor weather unless they're equipped and capable and go on an IFR flight plan, fly a lot more safely per mile than sober people drive in their cars. So why worry?

That sort of thinking and rationalizing can allay our fears enough to make our flying comfortable and enjoyable, and permit us to function with reasonable efficiency. But this is just the first layer. The fear is still with us, buried deep in our subconscious. Our next layer of accomplishment is to achieve a condition in which fear will never surface in an emergency and make us ineffective.

WINNING OVER FEAR

In a stressful situation, when our routine repression of fear breaks down, it appears that we can function effectively only by transcending fear. This concept has not been fully explored scientifically. I ran across it recently in a brilliant article by the late Emile Benoit, a professor at Columbia University. He was talking about the repression of fear of death in day-to-day living, and the need and evidence of transcendence of fear in emergencies. I think his hypotheses are relevant to our flying.

Transcending fear means forgetting oneself and charging in the direction of what appears to be the solution of the immediate problem. For instance, people who have received Carnegie Medals recall no sensation of fear. They simply transcended any thought of personal peril and thought only of how they could save a life.

UNSUNG HEROES

I think there is also a gleaning of truth about transcendence of fear to be found in a remarkable book recently published in England, *The 1,000 Day Battle,* by James Hoseason, an illustrated account of operations in Europe of the Eighth Air Force's 2nd Air Division, particularly its 448th bomb group and the other B-24 Liberator bomber units based in East Anglia's Waveny Valley.

Consider the realities which our airmen had to learn to live with. The English bombed at night when antiaircraft fire was less deadly and the bombers harder to find by enemy fighters. Our plan of operation, which the British had serious reservations about, was daylight bombing for greater pattern accuracy. The most important targets lay beyond fighter-escort range, deep in enemy territory. A hundred bombers out and only 75 back was a not uncommon experience. Even when the P-51 Mustang long-range fighters began to accompany them, B-24 losses averaged 10 percent per mission. It took an improbable 30 missions, and later 40, for a transfer back to the States.

Hoseason writes, "Formation flying allowed no evasion action, otherwise collisions would occur. So the bombers continued exactly along the track—the predictable track. It was up to the gunners to deal with the attackers. Only the gunners could redeem the situation, and they did their best." And in an account of a specific flight he continued, "There was just a single pass by enemy fighters. The P-51 'Little Friends' arrived on the scene promptly, having been alerted by the leader's call, and took on the Germans. The fighter pilots' air battle drifted to the south.

"Jonson considered the situation. In combat the individual vanishes. Men turn with absolute trust to one another; they need one another as they seldom do in time of peace. How important the thousand dreary routine things in the Army were. The drill, the saluting, the uniform, the very badge on your arm and breast all tend to identify you with a solid machine and build up a feeling of security and order. In moments of danger like this an airman turned to his mechanical habits and drew strength from them."

Later, "Navigators were training to be reliable record-keepers and consequently they became almost compulsive diary-keepers. Bill Verner, a Californian, navigator of Crew 31, wrote in his April diary: '24 APL. John A. McCure of 715 Sqdn., and his crew failed to return today. Begin to realize what they call heroism is only the act of a desperate man galvanized into action by his instinct for survival."

At this point let's think a minute about today's general-aviation pilot who loses the engine on takeoff in a single-engine airplane. Will he panic and spin it in a frenzied effort to get back to the airport? Or will he be able to transcend his fears and pursue the main chance, which is to get it down in the largest open space ahead within reach?

CONTROLLING PANIC

I am reminded of an incident in which airline captain Bob Buck was once involved. In a climbout on instruments at Milan one night, he lost an engine on his Connie. His flight engineer started chanting, "Oh Lord, oh Lord." Bob said to him, "Let's leave the Lord out of this and concentrate on doing it right." With that the engineer started thinking again and they got the engine secured properly, obtained a clearance and landed. This is not to be disrespectful of any-one of religious bent who finds solace and serenity in the air in faith.

Once, at Groton, Connecticut, Frank Nagel told me of an incident involving one of his Ercoupes. A local clergyman got into cloud at altitude and lost control of the airplane. Instead of the more common performance of pulling the wings off, he simply turned loose the wheel, got out his prayer book, and started reading it. It is hard to imagine what attitudes and airspeeds the airplane got into, and at some point it must have been inverted. When the minister finally got into the clear and landed back at Groton, he was black and blue from his chest to his knees from having slid back and forth under the seat belt. The prayer book and a strong and well-gentled airplane had saved him from what he might otherwise have done to the airplane, and to himself.

The question, of course, is whether transcendence of fear can be taught. Or, in plainer terms, whether pilots can be taught to keep their instinct for survival from causing them to panic rather than think in an emergency.

In the first airman training manual of the Signal Corps there was this sentence, in block letters: "NO PROPERLY TRAINED PILOT WILL EVER SPIN IN." That is probably true, but we have not learned how to do this. In these last 77 years of aviation history thousands of pilots have spun in, many of whom had had extensive spin training, many of them military, and many with instructors aboard during forced-landing practice. Since an airplane will enter and stay in a spin only with the help of a pilot, the problem is not the airplane but the pilot. Our spin-ins are due to a failure in our educational system.

When a spun-in airplane is found with its nose in the ground and its tail arched forward over the cabin, scorpion-like, it seems logical enough to conclude that the pilot obviously did not know enough about how to get out of a spin. And if anyone mentions that this particular pilot never had any spin training, this brings the crusaders out of the woodwork in droves. Here, at last, is a real chance to destroy evil. Not understanding what the problem really is, spin training is advocated.

Spin training? Starting safely at over 5000 feet above-ground, the student gets a spin demonstration. After that he knows what's coming when he does one, so there is no element of surprise. So far aboveground, he's not going to be greatly alarmed by the nose-down attitude of the airplane after spin entry or feel any urgency about a quick recovery. When he finds the nose has to be put even farther down to initiate a recovery, he may tense up a bit but he's able to do it.

To get the spin, with power off or reduced considerably, the nose is raised slowly above the horizon in straight flight. When the stall signals come—a shudder, or the sound of the airflow separating from over the top of the wing, or the nose starting to fall down and maybe a wing starting to drop—he gives it full back stick, full rudder one way or the other, and around and down she goes in autorotation. It is not an unpleasant sensation, and many pilots enjoy spinning. Unfortunately this training programs the pilot to think that what really gets a spin is full rudder at the stall. It leaves him unprepared mentally for the rudder-near-centered unintentional low-altitude stall/spin situation.

No one spins in from 5000 feet or more except test pilots who are unable to bail out. The consumer's spin-in problem starts below 1000 feet and at pattern altitudes in turns to crosswind, downwind, base, or final, or in a frenzied attempt to turn back after an engine failure on takeoff. What is overlooked in the search for an answer to the spin-in problem is the difference in pilots' reactions to an intentional spin at altitude and an unintentional stall/spin situation at low altitude, when they unexpectedly find a wing dropping and the nose going down and the ground starting to come up, fast. It is only instinctive to try to get the wing and the nose up, and this is what causes the airplane to spin in. The reaction is the same whether the pilot has had intentional spin training at altitude or not.

The answer as to how to get this millstone of ground-shyness from around the neck of pilots in our unintentional and unanticipated low-altitude stall/spin misadventures may be found someday in required and extensive use of a proper simulator. One in which a pilot could experience the visual effects of an incipient stall/spin at low altitude, both in straight and turning flight. We will not get the stall/spin problem in hand until a way is found to determine whether a pilot is too ground shy to make a proper correction in an incipient stall at low altitude and learn how to train this out of him.

THAT SINKING FEELING

As well as learning that we are our own worst enemy when it comes to sensing an incipient loss of control at low altitude, which may even have come about solely from a gust or updraft effect when we were flying too close to stalling angle of attack, we need to recognize that a suddenly developed high rate of sink can trigger the same response. Since we think in terms of elevators controlling altitude, our instinctive reaction is to pull back on the stick. If, it seems, back stick will give more altitude, it will also give less rate of sink.

Some of our airplanes, even if the speed is reduced only a few miles below normal approach speed, can develop a rapid increase in rate of sink. That's easily corrected by getting the speed back up to normal with the nose slightly lower, or adding more power, or both. The main consideration, though, is that the pilot first has to suppress his back-stick impulse and concentrate on flying the airplane out of the high sink rate. In this respect the sudden high rate of sink close to the ground is in the same category as the stall/spins at low altitude. In the sink cases the airplane may neither stall nor start to spin. It may simply land flat and hard, short of the airport. In both cases the cause is pilot rather than airplane flight characteristics—or at least if the pilot flight characteristics were correct, there'd have been a recovery.

FLYWHEEL EFFECT

What prompts this discussion of the flywheel effect is the recent reporting with some regularity about some small corporate jet coming in for a landing in gusty conditions, flaring, and then the next thing witnesses see is a wing rolling back and forth, a tip contracting the runway, and a disastrous cartwheeling of the airplane.

How could this be? Surely every pilot in command of a several-million-dollar airplane has had to prove his competence to someone able to judge it. Did he flare too high, hold it off too long, get into an incipient stall, and have to start fighting a tendency for the airplane to roll? In some cases the flare was said to have appeared both high and fast. Were stability characteristics of the airplane in the ground cushion a factor? Or was the cause of the problem the flywheel effect —the pilot's reaction to the inertia effect in a situation in which he needed to get a wing up quickly? Some of these jets have a quite high rate of roll once they get rolling.

Overcontrolling is something we all experience and overcome early in the game. The instructor says to get the nose up and we get it up too much. When told "not so high" or "not so much," we put it down too much. It may be a while before we start making smaller but adequate changes in control pressures.

The flywheel effect involves an inertia and kinetic-energy factor that is not a factor in rudder and elevator overcontrolling. It shows up principally in twin-engine airplanes and is a function of lateral distribution of mass. It really raises its head when you have large tiptanks or fuel cells virtually out to the wing tips.

If one has been flying principally singles or twins without tiptanks or far-out wing tanks, standard procedure to raise a wing is to give it the ailerons and, just before the wing reaches level, center the wheel. What with the damping in roll provided by the usually lightly wing-loaded down-going wing and not excessive lateral distribution of weight, the roll stops almost instantly when the ailerons are centered.

My first experience of the flywheel effect was in an early Cessna 310. Fifty gallons out there on each wing tip meant 300 pounds on each end of the barbell, plus, of course, 500 or 600 pounds on each side in the engine nacelles.

It was down at Winfield, Kansas, on a gusty, turbulent day. On final I got the 310 to rolling back and forth from right wing down maybe 30° to left wing down that amount, and I couldn't stop it. I knew how, but that didn't help. When I'd neutralize the ailerons as the wings went past level, the airplane would just keep on rolling, since I hadn't put on the brake soon enough. Actually the deflections of the wing were on the increase.

Finally my check pilot simply froze the wheel in neutral and gave it rudder as necessary to prevent any yaw; the wing came up and stopped level. The trick, of course, in picking up a wing in this situation is simply to neutralize the ailerons *before* the wing is level and it will roll onto or close to level, and then a small aileron correction will put it where you want it.

The Twin Comanches I flew for some years had two fuel tanks in each wing and 15 gallons in each tiptank, or 90 pounds out there. At times the flywheel effect could be felt with full tanks but it was really barely noticeable and did not bother me.

I have heard from pilots who went into four-engine aircraft with wings wet virtually to the tips that one of the first adjustments they had to make was getting used to the slowness with which the airplane started rolling in response to aileron use and the slowness with which it stopped rolling with the ailerons neutralized.

In an airplane in which the flywheel effect might occur, the pilot can easily take a look at what his timing ought to be if after a flare a wing should drop precipitately. At altitude just put a wing well down and then see how fast it can be raised and stopped precisely in a wings-level attitude. This makes it possible to judge the likelihood of an overcontrol reaction being triggered in such a situation, which a little practice would get in hand.

Somehow this all seems a bit redundant, but these cartwheel accidents do happen and it seems reasonable to assume that the pilot may not have been aware that in a showdown he would respond with an overcontrolling reaction. I'd at least rather believe this than that the accident was due to causes unknown.

In the end, when it comes to pilot characteristics, I think we need to think in terms not only of how to use our remarkable physical attributes to best advantage, but also what we need to add to them intellectually: we need to be sure of acquiring some reflex actions that are at first directly opposed to our instinctive reactions.

TOO GOOD TO BE TRUE

Recently I bought an hour's flight with a flight instructor whose students always pass their flight tests. He flew on the right side, and the purpose of the flight was to show me from start to finish exactly how he wanted his students to fly on their official flight tests.

He had a delicate touch and it was an impressively smooth and well-coordinated flight throughout. Straight as an arrow on the takeoff run, hand on the throttle. Lift-off at exactly the correct airspeed. Attitude constant, ball centered, in the climb. Here and all through the flight he was on the trim control rather constantly.

At altitude the first maneuver he showed me was the departure stall. Power back to 50 percent or less, nose raised quite slowly, and finally at the first sound of burble atop the wing and buffet of the tail surfaces, he put the nose down to level attitude, or maybe just a hair lower, gave it full throttle, and flew level and trimmed back to cruise speed.

Even our trainers are getting cleaner all the time, and I believe any instructor is a bit hard pressed to teach what needs to be taught about stalls in turns. If the power is left at cruise or higher and the turn isn't more than 45° bank, the new trainers seem determined to keep flying, round and round in level turns. The demonstration was with 50 percent or less power and made a turn-stall appear most certainly innocuous. With this power and maybe just a bit of climb in the turn, as the airplane began to shake and shiver and wobble just a bit, he raised the wing briskly, added full power smoothly, and away we flew back to cruising flight and trim.

And so on. Through the slow flight-maneuvers smoothly; correct power and trim and speed on downwind; 10° flaps and trim into a slightly nose-down turn onto base; a level turn onto final followed by a slight power reduction and then full flaps; trimmed precisely to full-flap approach speed and RPM around 1200, he was possibly overshooting just a bit when he closed the throttle and rounded out into a tail-low touchdown. The touchdown wasn't the softest I've ever seen, but just before touchdown a gust held us off for a moment. Even so, it would count as a good check-ride landing.

While impressed by the flight and able to see some needed improvements in my own activities, it still left me with a strong conviction that in today's training we do not have a proper balance between salesmanship and realism. Everything I saw implied that if flown just so, the last thing an airplane would ever do is bite. It's no doubt desirable to inspire confidence, but I do not believe that is enough. I realize that an early problem in instructing is often getting the new student past being so afraid of the airplane that he can't start learning. But somewhere along the line, toward the end, or even in a required hour's training after a license is granted, I wonder if a lot couldn't be gained by letting the student learn what can happen if he doesn't do everything right or everything doesn't keep working as expected.

The FAA, of course, prescribes the training curriculum. Their birth certificate mentioned responsibility for safety and also the fostering and development of general aviation. Success in bureaucratic management, just as in private enterprise, embraces the concept of growth. More flying, bigger bureau, new opportunities for advancement. From the beginning, FAA field personnel just as much as anyone else around the airports have been interested in making learning to fly sound and appear to be easy, which it is, and safe, which it isn't, or at least not safe enough in general aviation.

As at least one example of a safe-flight addition to the curriculum, I have always liked to see what a pilot does when the throttle is cut (at altitude, of course) in a maximum angle of climb attitude and airspeed and he is asked for a 180° turn-back as quickly as possible. It will often happen that with the bank started early and steep enough, the airplane will wind up with the nose well below the horizon and the pilot pulling so hard on the wheel that the nose starts going farther down rather than up as he wants it to. That's the split second in which it's possible to see when you've really made a pilot.

Someone, of course, will heckle, "Well, that's what we've been saying all along. You got to teach stalls and spins." But that's not what I'm talking about. Intentional stalls and hard-rudder spin training gave us a lot of years of plenty of spin-ins, and still would if this training were repeated. Because it doesn't bear on the problem, which is the unintentional stall, with the rudder centered or nearly so, at low altitude.

What I'm talking about is not intentional-spin training, but unintentional-stall and recovery training, which is the real-world situation we have to learn to handle, if, as, and when. And it's easy to duplicate in our training, but isn't.

THE 7° SOLUTION

Finally, a critical element in a pilot's acquired flight characteristics is a clear understanding of what a delicate thing angle-of-attack control is in slow flight. Angle of attack is the angle between the relative wind and the wing. The big question is where the relative wind is coming from, and we don't have the equivalent of the small wind sock atop a sailboat mast to tell us. Yet the magic angle for us in slow flight is only 7° or thereabout. For maximum performance we don't want any bigger angle than that between the relative wind and the wing. As far as attitude flying goes, we'll always find our 7° in our general-aviation piston aircraft, in a stabilized climbout or approach, between a nose-up angle of 10° in a full power climb, to proportionally less with partial power, to about 5° nose-down with no power. So at least we have these very positive and critical attitude limits to tell us where to look, and, for sure, where not to look, for a power-on or power-off attitude which will give us the superlative protection of a 7° angle of attack in a maximum-rate climb, slow level flight, or descent.

The next four chapters deal primarily with the operation of single-engine airplanes in the takeoff, approach, and landing phases of flight. Although they are applicable to twins in many respects (as long as both engines are operating normally), the twin on one engine is an utterly different machine and its characteristics and severe demands on the pilot are covered in Chapter X, "Engine-Out Twin-Engine Flying."

5

CLEARED FOR
TAKEOFF

Of course, in any airplane we just don't just get in and go. Or at least we are not supposed to. First there's a thorough preflight walk-around, then meticulous checking of each item on the takeoff checklist, and finally run-up as prescribed: these actions deal in known quantities and are bastions of confidence and safety for us.

BEFORE YOU GO: STOP AND *THINK*

But I do feel compelled to mention, from experience, that at our present evolutionary stage as pilots, some of us need to be ever mindful that somehow we often tend to become mesmerized as soon as we get within 50 feet of our airplanes. We look but do not think, and sometimes I feel that the second the wheels leave the runway, our schmarts depreciate instantly maybe as much as 50 percent—unless we make sure that we are concentrating not on ourselves, what we want to do, where we're going, or whatever else we may be daydreaming about, but on whether everything about the airplane and our management, so far, of the flight, is shipshape. Our lapses include everything from having to shut down an engine and get out and remove a chock from in front of the nosewheel, to finding our seat belt on the floor in the climbout's first bump, or having skipped carburetor heat check in the run-up, to having failed to remove the pitot's cover, to having taken off on a level-flight-only tank, and sometimes even waking up to the remembrance of having left our charts on the airport counter. Maybe we are to some extent victims of our jump-in-and-go automobile habits. When was the last time you checked the tires on your car for any serious cuts in the treads? If we think well enough on the ground, we won't have to do near as much of it in the air.

THE ELUSIVE CENTER LINE

Normally the takeoff run and the initial climb are simply a question of keeping it running along the center of the runway in a level attitude to a prescribed speed at which the nose will be elevated approximately 10° to climb attitude. Shortly afterward the airplane will lift off and for the next couple of minutes during the climb to pattern altitude, the pilot's job is simply to keep it on the runway heading and make whatever adjustments in attitude may be needed to keep the indicated airspeed on the maximum rate of climb speed.

This is, of course, an idealized description of a normal no-drift takeoff and initial climb in a tricycle fixed-gear trainer, but it does reflect the base line in both instructional and regulatory presumptions concerning this phase of flight. There has to be a starting place, a norm, a sort of theoretical standard of performance as a foundation for learning. But we all know that the instant we start moving, things start happening that modify, sometimes severely, the pure concepts of flight.

The airplane doesn't want to run straight along the runway, nor does it want to hold the runway heading in the climb. Right after lift-off we may find the airplane drifting sideways off the runway and wanting to touch back as a result of a premature rotation. If there's turbulence, the nose doesn't want to stay in what we consider a normal climb attitude. In some retractables, raising the gear calls for considerable re-trimming of the elevator, as is true of raising takeoff flaps. And so on. We are the sensory and servo system for our airplanes, and our competency revolves more around the early recognition of and attention to these variables to which the machine is subjected than it does around deft manipulation of the controls. Doing the right thing at the right time is more important than doing it smoothly. That can come later.

Variables? Our takeoff run—our spirited transition into the future—is a low-risk phase of operation from the personal standpoint, but it is surprising what-all can happen to an airplane just running along on the ground. In an average year, some 200 airplanes are substantially damaged in takeoff runs, and another 50 hit the fence or go into ditches in aborted takeoffs. In these 250 accidents, there may be 2 or 3 fatal accidents and 4 or 5 serious-injury cases. The balance involve no or only minor personal injuries. The usual official pronouncement is "loss of directional control." That, of course, is about all that could happen, but why does it happen? There are almost as many reasons as there are accidents. And a disproportionate percentage of these misadventures occur during both single and multiengine training on fields with relatively short and often narrow runways.

A taildragger pilot gets in his first tricycle-gear airplane, which has the nosewheel hooked to the rudder pedals, and, accustomed to using almost full rudder-pedal travel in the initial stages of his takeoff run, he puts the right foot-pedal to the wall, swerves violently, and goes off the runway, catching a wing tip and sometimes nosing over. "Ground/water loop/-swerve" becomes the official classification of the incident.

A tricycle-gear steerable-nosewheel pilot gets in his first taildragger, and unaccustomed to more than moderate initial rudder-pedal movement to keep it straight, goes off into a fast left turn, drags the wing tip on the outside of the turn, loses directional control, and sometimes goes over onto his back from full brake application on both wheels. (The Luscombe is especially distinguished in this category because of its awkward heel brakes, and I wonder if there's even one of these hardy perennials left that hasn't been on its back.)

A pilot stepping up from one of the properly docile trainers into more power may, with a good crosswind component from the left, not give it the boot that's necessary to keep it straight in the early stages of the takeoff run.

A pilot wanting to run it level for some extra speed in a crosswind takeoff may find the nosewheel steering in his tricycle-gear airplane suddenly so touchy that he overcontrols and swerves off a narrow runway.

Sometimes pilots use a combination of rudder pedal and toe brake for directional control, which, of course, lengthens the takeoff run. This catches up with them sometimes in the stress of a short-field takeoff.

Some of our tricycle airplanes have hind legs so short that their wing has an extra-high angle of attack in the takeoff run, which the pilot cannot reduce with forward wheel pressure because the nosewheel strut cannot be compressed enough to do the job. At a certain speed these airplanes lift off at their own discretion and the pilot's job becomes keeping them in the air at that speed while accelerating to maximum rate of climb speed of 10 MPH or so higher. This characteristic occasionally throws a pilot in a crosswind takeoff. Wanting extra speed before lift-off, he may give it full forward wheel, and as a result his powerful stabilator lifts the mains off the runway slightly while holding the nosewheel on the runway. Somebody in the FAA must have rolled a wheelbarrow at some time, because the very apt term "wheelbarrowing" was coined to describe the swerve and loss of directional control that a crosswind and especially a gust from the side can cause, when you run along with nothing except the nosewheel on the ground.

ROUGH/SOFT FIELD OPERATIONS

A paved-runway pilot inexperienced on rough unpaved runways may underestimate the effect a rough field can have on the length of the takeoff run. With a tail-wheel gear, he needs to avoid trying to pull it off because that extra back stick-pressure may mean a bump will give him a lively pitch-up before he has flying speed. He needs to get the tail up and keep the mains pounding the ground running level until it's ready to fly. With a tricycle, if a pilot tries to haul it off early on a rough field, the bumps may cause a pitch-down back onto the runway, and in subsequent oscillations a lot of extra runway will be used. Also, in each case, getting launched prematurely by a bump can mean a touchback with drift if there's a crosswind component, with a resultant ground loop in the tail-wheeler, or what looks like a ground loop in the tricycle, when the pilot tries to pick up the low wing with coordinated rudder and ailerons.

A pilot, especially a pavement-oriented one with soft-field takeoff experience limited to his flight test on a paved runway maybe a long time ago, will, when he encounters a soft field, or tall grass, or a snow-covered runway, revert to what was good practice in a side-by-side trainer. He may never have read the owner's manual in his present airplane, which is especially likely if he's a renter pilot of a single-engine airplane (they do half of all flying). Had he looked in the manual at the recommended short- or soft-field procedures, he might have read that while they wouldn't get him over a higher barrier, 10° flaps would shorten the takeoff run on a short field. For a soft-field takeoff, maybe 25° flaps are recommended. Or even 40° for the earliest lift-off, with the stick full back early in the run. But in this case with the specification that once airborne the airplane should be kept in level flight until it has accelerated enough to permit raising the flaps gradually to at least 25° without a touchback. On the other hand, there are airplanes in which you won't get the mains out of a soft field in a stick-back takeoff run. The Globe Swift, for instance, or the T-tail Lance if loaded.

These examples concern mostly low-time-in-type pilots operating off small airports, but come winter in snow country, the experience level of the pilots involved in accidents goes up quite a bit. These are the times at which we all learn that with not enough room to spare between the mains and low snowbanks, or between the wing tips and higher snowbanks alongside the runway and with the center line obscured or nonexistent, it is difficult to control the airplane's position laterally as precisely as in normal conditions. Again, "loss of directional control" with sometimes a postscript, "wing tip contacted snowbank," or "snow in windrows." (Even the big airline jets have had weathercocking incidents when they've taxied in crosswind conditions onto a sheet of ice on a runway covered by snow.)

And so on. There is a great variety of loss-of-directional-control incidents in takeoff runs due to the pilot's not concentrating on staying right with the airplane throughout the run and doing whatever is needed with the controls early. These examples may sound like student pilots to you, and many of them are, but that's not the whole story. Half of today's flying is done by pilots who fly no more than 50 hours a year (after all, that takes at least an afternoon a week if done regularly). Many of them, though, may have been flying a long time.

GETTING OFF THE GROUND

What do we need to be doing and monitoring and perceiving in our takeoff runs? What should be routine with us in each takeoff? There's no way to say what should be done specifically at any given instant in all airplanes, as no two makes or models fly alike. But it must surely be that in the millions of uneventful takeoff runs made each year, there is a common denominator of alertness and early awareness of any departure of the airplane from the straight and narrow.

Straight steering would naturally seem to be the principal element in maintaining directional control and it is, but the foundation of a good takeoff begins earlier, in the first part of the run.

Did the engine accelerate smoothly as the throttle was advanced? Is there normal thrust as evidenced by the force with which you're pushed against the seat back? Is the fixed-pitch-prop RPM normal after a second or two in the run? Or, are the RPM and manifold pressure normal at the same point? It is important to note the feel of normal thrust, because a constant-speed propeller can obscure a soft engine. If a takeoff run doesn't start with the usual gusto, it's a time to abort.

Is the humidity high today? High humidity, according to the Navy, can reduce power output of a normally aspirated piston engine by as much as 14 percent (the water vapor occupies space in the combustion chamber that would otherwise be filled with air for which fuel is being delivered for a normal fuel/air ratio).

In the four- and six-place airplanes with their greater CG travel range, trim setting for takeoff becomes more of an approximation. Does the nose seem to want to come up all by itself as the takeoff run progresses? If so, the trim setting for takeoff was too nose-up, and this can invite a premature lift-off. Or, does it take extra back-pressure on the control wheel to rotate for takeoff? If the trim setting was too nose-down shortly after lift-off, some nose-up trim will be needed to make the normal stick-force judgment of angle of attack available.

WHEN TO CHOP POWER

In grass, or when there's snow on the runway, or in soft-field situations, which account for most of the aborted-too-late takeoff runs, there's a warning signal that often comes on early. It is this: we find ourselves with the stick pretty well back, nose way up in the run, and we're about to lift off when the airplane stops accelerating. That, for sure, is the time to chop the power. Things *might* get better if we kept on—we might hit shallower snow, or less boggy surface, or shorter grass, but that's a long shot, and it can get worse. And, of course, there's the more obvious anchor to the windward in both short- and soft-field operations of having picked a point along the runway at which the airplane must be airborne or else the takeoff must be aborted.

So much for our takeoff-run problems in the rough. It's an odd situation. An airplane will almost take itself off. As a matter of fact many have. A pilot once hand-cranked a 60-HP two-place side-by-side Velie Monocoupe in Iowa on a farm field. No chocks, no one in the cockpit, throttle evidently well advanced. The airplane took off, departed the area, ran out of gas, and was found in the middle of a field five miles away without a scratch on it. Which should have told us then, and even now, that we must be overlooking something in pilot flight characteristics.

When it comes to our lowest-powered twins in the takeoff run, their pilots need to use a different standard for picking a point along the runway for chopping the power for an abort. Some of these twins, in which lift-off occurs at a lower speed than the best single-engine rate of climb speed, will not accelerate to that speed on one engine with the gear down without loss of altitude. That means the abort point for these pilots is well past the lift-off point. It will also be necessary to consider any obstruction handicap in terms of where the gear-up climb capability actually begins. All of which translates to the fact that if a twin-engine pilot really wants to have access to what his airplane has to offer in an engine failure on takeoff, he'd better keep the airplane on relatively long runways that have no hurdle at the end.

Otherwise, in a single or twin, rotating a bit early and letting the airplane fly itself off is a pleasant experience with a soft-enough gear. But with any prospect of a crosswind component, or loss of lift from an expiring gust in the run, my preference is to run up to a speed that will give a positive lift-off on rotation. That way any probability of a touchback is avoided. Right after such a lift-off there may be a sag in rate of climb as the airplane climbs out of the half-wing-span ground-cushion effect, which invites lowering the nose slightly for a speed increase to nearer maximum rate of climb speed.

This may not be the smoothest transition into the climb, but I think it is a more eye-on-the-ball procedure. Some spectacular angles of climb can be maintained for the first 50 to 75 feet after takeoff, because in ground effect a wing will fly at a higher angle of attack before stalling. And in the twin there's the additional boost from having the propellers blowing faster-moving air over each wing. But, single or twin, this is Russian roulette. In the twin, climbing on two below V_{mc} can mean a loss of directional control with an engine malfunction. In a single, why ever be caught with a power loss at low altitude climbing at a speed lower than is comfortable in a power-off approach (which may be imminent)?

OFF AND CLIMBING

You'll note from the table in the introduction that in the minute or so following lift-off in a flight, we are in one of the more critical phases of takeoffs and landings.

I do not have an official definition of "initial climb," but the accidents give a very precise meaning to the phrase. The action occurs between the moment of lift-off and the time at which, statistically, the pilot goes into cruise climb. In almost every initial-climb case the pilot hasn't even reached pattern altitude and is in the initial straight-ahead climb after lift-off.

The initial-climb statistical bin is quite a catchall. It includes each year maybe a dozen cases in which a pilot and friends, so potted you wonder how they found the airport, much less the airplane, took off from some small field, often at night and with limited visibility, and didn't even get to the first milepost. At other times, even in daylight, and sometimes with high-time pilots, you get a takeoff with virtually no ceiling or visibility, no instrument capability or equipment, and they lose control in the initial climb.

There are also a few cases involving stolen airplanes flown

by persons with little or no previous flying experience and, of course, if they do get off, they don't get very far. These accidents usually occur at night.

PRE-TAKEOFF PERFORMANCE CHECKS

A surprising number of the cases follow takeoffs on short fields followed by collision with trees, fences, or whatever else gets in the way. Often on larger fields the takeoff is from an intersection. Sometimes the field may not be so short, but the density altitude is high. It makes you wonder why the pilot didn't look in the owner's manual at the appropriate charts. But even then he might not get warning enough because those charts start with standard-air sea-level performance on a dry, level, paved runway with no allowance for turbulence. Even though the graphs show the performance for higher density-altitudes, the specifications for dry, level, paved runway with no allowance for turbulence still apply. Obviously, the figures are not too good a guide if the field is sod and rough, or if the run will be a bit uphill, or if there's high grass or snow. Or if there's a combination of these things. Where there's any reason to think getting out may be a problem, certainly a responsible pilot would first make a solo flight to see whether he wanted the additional load of one or more passengers and possibly baggage. There's nothing sissy about taking your passengers out of a tight field one at a time to the nearest larger airport.

There are also cases each year, involving surprisingly quite a few high-time pilots, who take off overloaded, or after a prolonged run on a soft field or in snow, or in a high-density-altitude environment and hit trees or wires or else try to turn to avoid the collision and go in.

There are a few cases regularly in which right after lift-off, a pilot realizes he's in angle-of-climb or even rate-of-climb trouble. Thinking that putting the flaps down will give him more lift, he lowers the flaps. Unless flaps are prescribed for takeoff, putting them down flattens the angle of climb because flaps create drag and use up much of the power otherwise available for climb. Flaps were invented to lower landing speeds. Coincidentally, their drag-producing ability gives better approach-path control. If they are put down with the nose held up in a normal climb attitude, there may be a momentary increase in rate of climb, but it will be followed immediately by a speed loss, and after that, if the pilot raises the flaps without first trading some altitude to get back to the proper flaps-up speed, he's in real trouble at low altitude.

In climb troubles in high-density-altitude situations, usually with a full load, putting the flaps down in the climbout where flaps are not recommended for takeoff often results in the airplane settling back to the ground outside the airport.

In balked landings sometimes flaps are raised abruptly with the airspeed below flaps-up stall speed, and a spin-in follows.

Each year's figures include a half dozen or so cases involving collision with obstructions or loss of control in the initial climb because of frost or snow on top of the wings.

POWER PROBLEMS

Surprisingly, almost a third of the initial-climb cases have engine failure or malfunction as a first cause. In this third there are usually some 50 twin-engine cases involving loss of an engine on takeoff, and these are quite frequently serious accidents.

The one third of initial-climb cases described as "engine"

accidents include, of course, taking off on an empty or near empty tank. Or on one tank when the placard says to take off only on both tanks. Or mismanagement of the fuel system with an aux-tank takeoff or failure to use the boost pump as prescribed. Water in the gas is also often listed as a factor. But in a surprising number of cases the engine failure or malfunction is listed as due to causes unknown. This label is a bit confusing because engines just don't quit without cause, and certainly structural or lubrication failures are a rarity.

After an accident the condition of the propeller will indicate whether power was being developed at the time of impact, but that doesn't tell whether the pilot may have closed the throttle of an operating engine just before impact. Carburetor ice is difficult to detect after an accident, and this malady starts with a partial loss of power. A vapor lock is even more difficult to determine. Or a clogged fuel-tank vent might not be too evident.

I often wonder, when a pilot reports loss of power in an initial-climb case, whether sometimes because he couldn't climb at the rate or angle he wanted to or needed to, he assumed this meant a power loss. When automobiles had brakes on only two wheels, drivers in collision cases often claimed their brakes had failed. Actually the brakes had often worked normally but just failed to stop the car in time.

At any rate, in these cases attributable to engine failure or malfunction in the initial climb, if the pilots don't hit something going out of the field, they often settle back to the ground outside the airport. And hit relatively flat. Which sounds a confidence-inspiring note because it proves again and again that if an airplane can be gotten into even a small fairly smooth area at least not entirely out of control, the personal risk can be minor. These cases not only speak well of the stability and control characteristics of present-day airplanes in slow flight, but are also a credit to the pilots for at least not having come in close enough to a stall to lose control of the aircraft. In looking for a still better record, my question is whether these hard landings indicate a movement of the ball from the airplane- to the pilot-characteristics court.

THE STALL/MUSH SYNDROME

Even without a partial power loss in the climbout it is possible to fly too slowly in trying to get more climb. At some point, because of turbulence, or density altitude, or an overload, or a downdraft, flying too slowly can bring on a descent rather than a continuing rate of climb. If this happens, it is instinctive for a pilot to keep increasing back pressure on the stick in an effort to stop the rate of descent, and this can lead to a hard and in some instances too hard substantial-damage landing. Or, in this too-slow situation, still further effort to slow the rate of descent with a higher angle of attack can result in a last-minute uncontrollable nose-down from a gradient-wind or wind-shear effect.

Then there's the most serious pilot reaction of all in these "stall/mush" accidents when a pilot tries a desperate turn back to the airport and spins the airplane. As far as airspeed is concerned, a tight turn is just like dropping anchor, as Alex Henshaw, the famous Spitfire test pilot of World War II, once put it. I'd add, especially in slow flight, which is the worst time of all to attempt to maneuver excessively.

But the main thing, in the mush cases, is that the pilot's reaction to an increasing or high rate of descent close to the ground is to increase the angle of attack of the wing. This is a time when he needs to be thinking in terms of maintaining last-minute flare power by keeping the angle of attack or airspeed as they are in a normal approach.

Do you remember the first time a jet airliner ever made a dead-stick landing? In the Atlanta area? Due to hail ingestion, both engines had stopped and they could not be restarted. The captain's windshield was opaque from the hail damage, but the copilot's side was clear and he made the landing, on a country road.

Can you imagine what a high rate of descent it must have taken to maintain a proper approach speed and how critical it was to start the flare at just the right elevation and rate?

But they got it on the ground under control. The last words on the flight recorder tape were the captain saying, "Don't stall it. Don't stall it." He knew.

Unfortunately one wing cut through a roadside filling station and the other wing tip shortly afterward dug into a roadside embankment. The crew and a number of passengers were lost, but it would have been a total without that skillful touchdown. The first requisites of a forced landing are self-control and angle of attack.

In general aviation most of them are successful.

6

THE TRAFFIC PATTERN AND CIRCLING

This statistical phase of our flying starts with the airplane having been leveled off on completion of its initial climb to pattern altitude and ends with it headed in on final approach. While the accident frequency is not so high in this category, it is clearly one that needs careful study because of its high incidence of spin-ins.

It doesn't happen often, but there's a very easily avoidable trap right at the start. The principle involved is going into a fairly steep turn with a low entry speed in our 90° turn onto the crosswind leg of the pattern.

In these cases the pilot levels off at pattern altitude with little more than maximum rate of climb speed, or in other words 7° angle of attack, and immediately racks it into a 45° bank. From the previous G-load in the turns table, this can raise the airplane's stall speed often 10 MPH, and if the turn is overtightened a bit to keep it level, the stall speed can go up even 15 MPH. This can mean the wing is then flying at an angle of attack within only a few degrees of stalling angle. This might not be so bad in calm conditions or in a steady wind, but if there's wind it's almost never steady and turbulence- and gust-free at pattern altitude.

AVOIDING GUST-INDUCED STALLS

On windy days it's easy to get caught in the turn to crosswind and be handed a stall recovery at low-altitude crisis when a gust hits the bottom of the 45°-banked wing as the airplane nears the crosswind leg heading. At this juncture a gust can increase the angle of attack well past the wing's 16° or so stalling angle of attack. The cure is simple. When it's really windy or gusty, stay out of harm's reach simply by increasing the entry speed a bit in this turn to crosswind and keeping the bank shallow, say no more than 15°.

The turn from crosswind to downwind seldom occurs with as low an entry speed as the turn onto crosswind, but to some degree it is still subject in its early stages to the just-mentioned gust effects on angle of attack if the bank is steep. And, of course, in the steep, level turn extra airspeed is lost. On a gusty downwind heading reached with too low airspeed, there can be a noticeable tendency for the airplane to lose altitude as it flies into the backside of gusts, and raising the nose is a common reaction, which can result in an interval of flying closer to stalling angle of attack than one realizes. While the turn to the downwind leg is possibly the least critical of the four accident-producing pattern turns, a bit of extra speed and shallow banks on the wild and windy days are still good investments.

Once on downwind not much ever happens. But the pilot can be mighty busy. Anything that can be done to keep the necessary chores from bunching up helps a lot. Our cockpit activities often take place against a distracting and even confusing communications and frequency-switching background.

Coming in from a cross-country or area flight for a downwind pattern entry, we need to lose some altitude and enter the pattern with the speed we want to maintain on downwind. The next step is easier if we know what power setting will give us that speed in level flight, gear and maybe first notch flaps down on downwind. Some airplanes want to nose down when the gear is lowered and nose up when flaps are lowered, and we have to anticipate and counter this with increased backpressure on the control wheel followed by increased forward pressure. Some pilots prefer to hold the nose in a level attitude as the gear goes down and then trim the load off the wheel. Then when they lower approach flaps, they again hold the nose level and trim the other way. This trim exercise can be reduced to only one, or in some airplanes no extra trim change, simply by holding the nose fixed as the gear goes down and keeping it fixed as flaps are then lowered after the gear is down and locked. In many airplanes this requires little retrimming and in some airplanes none.

PRE-LANDING CHECK AND DOUBLE-CHECK

Fuel selector on the prescribed tank or tanks for approach? Cowl flaps open? Known or verified ground frequency set on the second receiver? Carburetor heat? And whatever else is on the landing checklist. Plus looking for traffic, even at a controlled field. These are some of the chores that complicate our lives in preparation for our approach.

Toward the end of our downwind leg we keep checking on the end of the runway to see when a turn to base leg would be proper. At this point we start concentrating almost entirely on the niceties of the approach phase, and give little thought to the fact that the bulk of the misadventures in preparation for an approach occur in the turn onto base leg and the turn onto final. On downwind with a good tailwind, we can get in in a hurry to turn to base, or the tower may ask for a short approach, and we rack it into a too-steep turn, forgetting that with extra G-load, increase in angle of attack far outpaces deterioration in airspeed.

WHERE APPROACH PLANNING PAYS OFF

Our next turn onto final approach course causes even more trouble than the turn onto base. An overshot runway center line engenders an overpowering impulse to make a too steep, too tight correcting turn and we lose control.

If a steep turn has to be made at these critical junctures, putting the nose down at least halfway to the first below-level mark on the artificial horizon before even starting the roll into the turn, and leaving it down there during the turning interval, would be good insurance. This gives us more power, from gravity, to offset the high increase in drag during the turn, more power than the engine may be capable of delivering at the time.

CIRCLING: A CAUTIONARY TALE

What about the "circling" part of this chapter title? Here's a typical scenario. The weather's perfect, we've had a delightful flight, and five miles out we spot the hangar on the airport. No answer on unicom about the active, so we decide to circle the field to look at the wind sock and also to check on anything that shouldn't be on the runway. Deer, dogs, or cattle or whatever may have strayed onto the runway. Even hardware may have fallen off airplanes or maintenance trucks.

We're new at flying and no one has ever told us that loss of control in turning flight at low altitude is where nearly all spin-ins start. So to keep our turn more or less over the airport, we bank 45° or a bit more and tighten the turn as necessary to hold altitude. Thinking in rudder-turn terms, we may even be holding some bottom rudder and opposite aileron to keep the bank from increasing, which puts still another wire across the flight path.

It's always worth repeating: as the airplane started to roll out the bottom of the turn, if the pilot had moved the wheel forward even only an inch, the ailerons would have started working and the pilot could have rolled out of the turn and continued his flying career. Instead, the instinctive reaction is first to turn the wheel hard over, away from the direction of turn, in a desperate attempt to unbank, and simultaneously, as the nose corkscrews down, to pull the wheel all the way back in an effort to get the nose pointed up. This is the stall/spin millstone of general aviation; suddenly, unexpectedly, and at low altitude.

Sure, before entering a pattern, look the small country airport over, carefully. But limit the bank to 30° in smooth air or 15° in turbulence, and keep the power up. Airplanes don't spin themselves. The stick has to be held back.

MIDAIRS

Finally, there are midair collisions in the pattern/circling phase. For convenience let's use them as a transition between this and the next chapter even though most of them occur in our final approach.

Most of the collisions occur around small, less busy airports, where we tend to assume there's little traffic. The traffic, of course, need be only one other airplane. Pilots can't maintain a continuous maximum alert in all phases of their flying, and they don't need to. But most of our collisions involve one airplane on final and the other turning final or overtaking on final. So final is a place to do a lot of looking. This is a bad time, unfortunately, because on downwind, base, and final, we're intent on judging our approach path, which means repeated seconds-long gazes at our hoped-for touchdown spot on the runway.

VIGILANCE IN THE PATTERN

Maintaining the needed scan for traffic is also affected by habit: en route we do little looking for other aircraft because no matter how much looking we do, we seldom see another aircraft that is significant traffic to us. In time we look less and less en route. And probably carry a lot of this habit into the traffic pattern.

For the see-and-be-seen concept to work we need to look around more than we do in the pattern. We can avail ourselves of the see-and-be-seen part by making proper town-/airport blind broadcasts on the 122.8 nontower field party line announcing that we're entering left or right downwind for runway X, turning base close; or far out for runway X, and turning final for runway X. This procedure has prevented many a collision with a hitherto unseen airplane at an uncontrolled field.

7

CLEARED TO LAND

In an average year a surprising 275 airplanes on final VFR do not make it to the airport. That's a lot of airplanes and a big slice of the 420 reportable accidents indicated in the "on final" category listed in the introduction. In a way it's puzzling. Once safely on final we have only to keep it headed toward the airport at a proper approach speed and rate of descent. This should be the easy, downhill part of our activities. What happens?

Many of the cases reflect the efforts of aircraft owners to get off the beaten path to get maximum usefulness out of their machines. There's no official figure, but many on-final VFR accidents happen with the airplane approaching a small out-of-the-way airport, which often means clearing close-in approach obstructions. Trees, wires, power lines, poles, fences, mounds, all of which tend to be small-airport standard equipment. So the problem is a case of collision with objects other than the ground, from undershooting the approach, not seeing the obstacles, or getting into a downdraft and not bringing up the power soon enough. It is also significant that in many of these cases the pilot learned to fly on a big airport, and among the many invaluable judgments of a small-airport habitué he has missed learning is the knowledge of when and how often it is best to go around for another better-organized shot at the field.

THE LONG AND SHORT ENGINE-OUT APPROACH

The on-final VFR figures also include twins coming in on one engine. The pilot undershoots only to find that with the gear and flaps down, the airplane cannot maintain altitude on one engine. Which suggests that even with both engines operative, dragged-in approaches in twins mean getting out on a limb because if an engine is lost, it is impossible to make it to the field on the other one. The best practice would seem to be to keep all twin-engine approaches slightly on the high side, since no twin is going to glide very far with gear and flaps down and all power off. Fear of overshooting in landing configuration in an engine-out approach, since that's likely to be even worse than undershooting, may be what leads to more under- than overshot engine-out approaches.

Or maybe the scales are tipped toward undershooting because pilots' everyday practice of dragging it in on two, causes them not to appreciate fully the extra glide-path control they have with only one engine operating if they are not accustomed to slightly steeper, lower-power twin-engine approaches than most pilots execute.

A factor, too, in the everyday dragged-in approaches in twins may be pride. A go-around on both engines in a twin does not suggest the degree of skill which twin-engine pilots enjoy having acceded to them. So they purposely slightly undershoot every approach in a trade-off of pride for false pride. But, to give twin-engine pilots their due, they are, on all except large airports, under a lot of pressure to land short because their touchdown speeds are in the 70- to 80-knot range.

Pilots, of course, also undershoot approaches in single-engine airplanes with everything working normally and get into the same trap twin-engine pilots can work their way into with everything working normally, except the pilot. In these cases the airspeed gets too low and the airplane develops a disturbing increase in sink rate. Once the airplane is on the backside of the lift curve, adding even full power a bit late isn't going to change things immediately, and the pilot can get suckered into raising the nose in an effort to slow the sink rate. Which may only increase it. These are the commonly serious stall/mush accidents resulting not so often in a stall/-spin tag, but a hard landing short of the airport.

The fact that they hit flat rather than go out of control as they more often did in earlier days is a tribute to our aircrafts' improved controllability and stall resistance. But an airplane can hit just so hard, flat, for an accident to be survivable. Occasionally they hit hard enough for the gear, in the low wings, to rupture the fuel cells with a subsequent fire. In more than half the cases of this type of too hard landing, the occupants exit with minor or no injuries.

SINK OR SPIN

The thing to remember is that on a VFR final a sudden big increase in sink rate close to the ground often triggers the same pilot reaction that an unintentional incipient spin entry does, namely to get that nose up. The only procedure that will cure the situation quickly is full power and lower angle of attack, even though the latter may mean trading some precious altitude for less drag and a faster increase in airspeed. Or at least I think this is the best procedure in our piston twins.

You should know, however, that the FAA has recently issued an advisory circular to the airlines that a sudden loss of airspeed on final from wind shear or a downdraft should be countered by raising the nose until the stick-shaker stall-warning signal comes on and adding full power. A simulator, of course, can be programmed for aircraft performance and downdrafts or wind-shear effects of any selected amount. There is a case of a 727 airline crew with this type of simulator training having arrested a high sink rate starting at 500 feet above ground level with a loss of only 125 feet in altitude by using full power with the stick shaker continuing to indicate operation in a near stall. But the question is how deep is the ocean and how high is the sky. What are the maximum sink rates that nature can provide on final? The jets, even in approach configuration, may have a higher rate of climb with full power than most of these natural phenomena. In our general-aviation twins I don't think we do, and it would be unfortunate if jet procedures were impressed upon the pilots of our much lower-performance aircraft. We can't lower the nose very much as we add full power, but I can't buy raising it. Also, with our piston engines, we can accelerate faster than the big jets can.

KNOW YOUR MACHINE, CLEAN AND DIRTY

Still another factor in undershot approaches is from pilots not having investigated at altitude how much power it takes to fly their airplanes level, gear and flaps down. In some airplanes it takes a lot of power. On a hot day with a heavy airplane in an undershot approach, we can come up too slowly on the power and get so near a stall that even full power won't do the job. There have been cases in twins in which even with both engines working normally the airplane landed, mushing, short of the airport. Twin or single, if a heavily flapped airplane has a quite low rate of climb in landing configuration, bringing it in on a day with scrambled air on final calls for less than full flaps and higher-than-normal approach speed. As well as being on full alert for the need for early and adequate power increases if the airplane starts sinking below the proper approach path.

POWER/TRIM ON APPROACH

Still another factor in the mush cases resulting in landings short of the airport are the trim/power-change characteristics of our airplanes. This is something every pilot needs to explore thoroughly at altitude in his airplane. Our high-performance airplanes today are designed to be strong willed about seeking whatever attitude it takes to regain the airspeed for which the trim tab is set, and also making whatever attitude adjustments are necessary to maintain that airspeed. Trimmed to a cruise speed you'll find that in nearly all of them it will take quite a lot of forward wheel to fly 5 knots faster, and also a lot of back pressure to fly 5 knots more slowly. For this reason we become accustomed to keeping after the trim tab to eliminate having to hold forward or back pressure to hold a desired airspeed.

This strong speed-keeping and, on occasion, speed-seeking stability characteristic of an airplane is a virtue. It reduces the frequency of structural failure from someone pulling back too much on the wheel in a dive. If it takes a strong pull, the pilot has time to think about what he's doing to the airplane. But these stability characteristics, plus downspring effects, require that we learn to anticipate their effect in approaches and landings when we make large power changes.

In most of our current production airplanes, in the usual power-on approach with the airplane trimmed for close to the proper approach speed, adding or reducing power the usual corrective amount to stay on our desired approach path requires little forward or back pressure to override any pitch-up or -down tendency. The wing and tail are flying at approximately the same airspeed.

There is a situation, though, say in forward CG, in which, if one is trimmed for a power-off glide at a proper approach speed, adding full power will require a lot of forward pressure on the wheel to prevent a lively pitch-up of the nose. In some airplanes it can be about all one can manage with one hand on the wheel and the other on the throttle. The reason for this is that, trimmed for the power-off glide, the trim tab is way down as necessary to offset the aerodynamic and downspring moments. When we go to full power, the airflow over the tail may jump abruptly to 10 knots more than it is over the wing, and up the nose goes. In such a situation we can't afford to be genteel and let the throttle take over attitude control. In an approach, and trimmed for a power-off glide, flaps down, some aircraft will stall if full power is applied and the pitch is left alone. This trim-power-change characteristic could be a factor in an undershot approach starting with power-off approach-speed trim setting. An airplane that seems overeager to climb can make a pilot think he's been overlooking the real capabilities of the airplane.

On the other hand we have to learn that with some of our only moderately flapped fixed-gear tricycles, what with the streamlining these gears are now getting and with no downsprings, the more limited power-change effects are helpful, but adding too much power in an undershot approach can give us an unwanted extra 10 knots across the fence.

The on-final VFR files also include the members of the Out of Gas Club, and associate members who mismanage the fuel system on final: those who select one tank (near empty) instead of both at the start of the approach, or both instead of fullest tank, or leave the fuel pump off, or otherwise omit what the manual calls for about the proper fuel-selection procedure on approach.

There are more landings short of the airport attributable to engine failure or malfunction than you might expect. But when a "from causes unknown" tag is added, it is hard to believe, since the engine had no structural problem, that the pilot was doing everything right. Did he go to full rich before starting down, or use carburetor heat properly, or did he pull the mixture control thinking it was the carburetor heat control?

The most frequent closing comment on the on-final VFR cases is "failed to maintain flying speed." That's hindsight. The more immediate question is how do we determine proper flying speed and how do we maintain it with precision.

VFR GO-AROUNDS

To the 275 airplanes that don't get to the airport in VFR approaches, we have to add 110 for those times when the pilot overshoots and has an unsuccessful go-around.

What happens? Again, the small airport is more often than not the scene of these misadventures. After coming in high and fast and floating halfway along the runway, flaps down, we give it the gun and pull the nose up. Starting with maximum rate of climb speed and the airplane in normal takeoff configuration and full power, the climb path is about 8° at most. With the flaps down it is normally only a fraction of that, and if there are trees to be cleared in the go-around, we often collide with them. Sometimes we even stall and lose control of the airplane in our belated effort to get back up for another shot. Sometimes, too, with the trees barely cleared and the nose high and the airspeed lower than flaps-up stalling speed, we raise the flaps abruptly and lose control. These, of course, are the more serious cases. At other times the trim/high-power situation as just discussed can be a factor. Engine failure or malfunction from causes unknown, as in the on-final VFR cases, is often listed as a factor and possibly for the same reasons.

TOO HIGH, TOO FAST

What happens doesn't always suggest where to look for the cure. Why do we overshoot? The most obvious cause is starting down in the approach too late, too close to the airport, and trying to get down by putting the nose down. This, of course, adds miles to our proper approach speed, and in our high-performance airplanes, an extra 10 knots after the flare can move the touchdown point in still air hundreds of feet farther along the runway. But it's not always as simple as a last-minute high-dive at the airport.

Often the airplane is at exactly the right altitude at the right place along the approach path and at the right airspeed when the overshooting problem presents itself. We fly into a strong updraft and with no change in pitch attitude start moving horizontally or at times even gain altitude.

What countermeasures are available to us? A lot depends on the flight characteristics of the airplane as well as the pilot's ingenuity. In an earlier day the airplanes were less complicated and had a lot more drag than they do today. At that time the answer to an updraft was either to sideslip, or slow down for a higher sink rate, or make some S-turns. This still applies to many of the older airplanes that constitute such a large part of the general-aviation fleet.

The sideslip is a powerful get-ridder of altitude, but with much in the way of flaps hanging down, it is easy to run out of ideas trying to sideslip with the nose down where it has to be to maintain proper airspeed. Also, some of our airplanes carry placards against sideslips with flaps down. Sideslips, you might say, have had their day.

Slow down to get more sink rate? This one calls for caution, of course, and also familiarity with the airplane's slow-flight rates of sink. It is highly effective in some airplanes and not in others. For instance, in a Bellanca Viking, for some odd reason, slow to just a few miles less than normal approach speed and the airplane develops a startling increase in rate of descent or sink. This is a very useful countermeasure for a Viking owner who encounters an updraft on final, provided he remembers that the way to get out of the high sink rate is to get the nose down rather than by just adding power.

In other airplanes, especially those that would not be considered overflapped by some pilots, I've found that slowing a bit to offset an updraft effect seems to make the airplane balloon more than ever. My most recent experience of this kind was in a new Cessna 152-II, which now has only 30° flap travel instead of the previous 40°. Slowing a bit in a strong updraft, I went from a low rate of sink to a lively increase in altitude. Going back to normal approach speed seemed the lesser of two evils, and the touchdown was still within runway touchdown-point limits.

Possibly more to the point: a few years back, flying in a Cessna 182, I discovered that when the airplane was flown, with full flaps, about 10 miles slower than normal full-flap approach speed, the rate of descent was less than in a normal approach. So that particular model, with its high-lift flaps, wouldn't gain by slowing up in an updraft. I find, however, that in a 1981 Skylane I've been flying, it doesn't take much slowing in an approach to get a helpful increase in rate of sink.

If a sideslip or slower flight in an updraft doesn't work, not much is left other than a mild bit of S-turning if there's enough distance and altitude to end up on track. In heavily flapped airplanes, power off, it is possible to nose down quite a bit in an updraft without getting an unmanageable increase in approach speed.

And, it's heresy, for sure, but even in a clean airplane, nosing down rather steeply to get rid of clearly excessive altitude works if it is done soon enough. Even a clean airplane, once the excess altitude is gotten rid of in this manner, will slow rather quickly back to proper approach speed if held nose level, power off. I've told as many new pilots as anyone has, "Don't ever dive at the airport," and that's still a valid stock admonition close in. But if you get at it soon enough, it's something else again. I learned this, eyebrows up, mouth shut, once from an associate, Ken Lester, who'd wound up World War II as a B-29 instructor. I was checking him out in my Culver Cadet, which had no flaps and little extra drag with its two small landing-gear legs down.

On his first approach, across Esso's mile-square refinery at Linden, New Jersey, Ken hadn't glided more than a few seconds on final, power off and at the proper approach speed, when he saw he was in trouble. He put the nose way down, the speed went way up, he leveled off at just the right distance out, and we crossed the fence with exactly the right airspeed and altitude for a normal short landing. But if it isn't done soon enough, forget it. And never on a license test or biennial review. This one's for when nobody is looking and you don't want to have to go around. And when it's warm enough that the engine won't suffer thermal shock in the descent.

THE APPROACH PATH

While updrafts are a good alibi for go-arounds, it is not likely that they are the cause of most go-arounds. Lack of proper approach-path control is a much more common cause. But, again, that's hindsight. In order to get the accessory before the fact, we need to know how we can tell when we're on a proper approach path. The statistics suggest there hasn't been a lot of progress in this part of the art of flying.

Once upon a time, in order to get a private license, the following maneuver was necessary: headed in on final, at 500 feet above runway level we had to cut the throttle at the point at which we thought it necessary in order to land beyond by not more than 75 feet a line across the runway. No sideslipping allowed, nor S-turns, nor (in a bind) more than one short clearing of the engine from 500 feet on down.

You can imagine what often happened. If we got into an updraft and landed beyond the 75-foot limit, or got into a downdraft and landed short of the line, this was supposed to indicate a lack of judgment of the gliding radius of our airplane. Also, and an allegedly more serious shortcoming, it indicated a lack of proper feel for the aircraft. A change in wind velocity on final could, of course, give us under- or overshoots. We also sometimes closed the throttle too soon or too late and got tarred with the same brush. So it was often a case of having another try a month later beneath a kinder sky.

Nobody explained what the inner secret of proper feel for the airplane was. Airspeed indicators were not required, and few airplanes had them. We had to rely on feel to judge the proper approach speed. Feel for the responsiveness of the rudder, ailerons, and elevator with the airplane flying at a proper approach speed. And most especially, feel for the responsiveness of the elevator, since it also provided a backup signal of variations in G-load.

Too slow in the approach, and movement of the controls would produce only sluggish responses. Sloppy controls was the phrase. Too fast in the approach, and control movements would cause clearly too lively responses, especially in the pitch response of the elevator. We learned the feel of control pressures and control responsiveness indicative of a proper approach speed.

"FEEL" AT THE FLARE

I think control-feel is still important in the touchdown phase. We use our airspeed indicators until the flare is started. From there on, so close to the ground, we can't be looking elsewhere than over or around the nose. Besides, the airspeed indicator's low speed, high angle-of-attack readings are so erroneous as to be meaningless. In the last seconds before touchdown, feel again becomes our flying-speed, or more correctly, our angle-of-attack indicator. Or indicator of remaining lift available. At least in tail-low landing operations.

The power-off approach requirement of earlier days did not reflect a neurosis about engine failure. One wasn't needed. They quit often enough (the OX-5 about every 50 hours with rocker-arm malfunction) for it to be good practice to make every landing a forced landing. The neurosis sometimes developed later in the realization that, aside from atmospheric effects, about half the time it was necessary to drag it in with power or else sideslip to avoid an overshoot.

Times, of course, have changed. Engines seldom quit of their own accord. We now teach a proper power setting and speed for the downwind leg, and from there on to base and the final throttling back, gradually enough to keep the RPM or manifold pressure from increasing because of the denser air we're letting down into, which would otherwise flatten our approach path. In short, power approaches are the norm.

With power approaches our over- and undershooting problems should have vanished: a little high on final, less power; a little low, a slight increase in power—such adjustments should get us there. But they haven't. Why doesn't this seemingly can't-miss system work better than it does? It doesn't because it leaves us plagued with the age-old question of how we tell when we're getting too high or too low in an approach.

In many of our attempted "positive control" power approaches we are a little late in discerning that we're getting on the high side. We wind up pulling back hard on the by-then closed throttle and not getting down more steeply. We are also often late in reducing power after increasing it to offset the effects of a downdraft. In consequence our overshot power-approaches often wind up as power-off approaches with all their limitations on approach-path control. In short, in both power-on and power-off approaches, we have the same difficulties in visualizing an extension of our flight path to a point on the runway over which we'll commence our round-out. The only way we could be sure of never overshooting a power-on approach would be to operate entirely on the undershoot side and drag it in with power each time. I think to some degree most of us tend to operate this way, to avoid long landings or go-arounds, but until we get 100-percent reliable engines, we need to continue to seek better judgment of and control of glide path.

EYE-BALL THE NUMBERS

The most useful theory to me is this: on downwind as we go past the end of the runway, and on base, and on final, we keep glancing at the numbers on the runway and call on our memory circuit to tell us if we're looking down at them at an angle which from past experience gave us a glide path within the gliding radius of the airplane. Our glances on downwind tell us about when to turn base, and on base whether to start turning early or not. Our downward sighting angle from pattern altitude gives us the only approximation we can have of whether we can make the runway with normal approach speed and power setting, or if, with the base closer and the sighting angle steeper, we can make it power-off.

If the angle to the numbers anywhere along the approach path is shallower than our experience says it should be, we know we're on the undershooting side and it will take more power to get in. If the angle to the numbers gets steeper than it should be, we know we're on the overshooting side and will have to use less power to correct or else do something to increase our rate of sink if power reduction doesn't do the job.

Thus, on the way in on final, we check at regular intervals and make power corrections as needed and hold our airspeed. We use our memory circuit and the eye's micro-capability in judging angles, and with precise speed control, our judgment of where we're really going becomes almost a subconscious activity.

Meanwhile there have been great improvements in what's available for better glide-path control. Some of the more heavily flapped airplanes tend more to go where they're pointed, especially the twins with low power settings. With heavier wing loadings in the singles, better sink-rate adjustments are available with power changes. And, finally, the designers are beginning to notice that maybe the glider people have something with their clean-as-a-whistle machines and spoilers for glide-path control. All their approaches are power off and they not only land on the spot but slide to parking slots on the line.

Otherwise, the art part—the ability to project ahead of us a continuation of our flight path in a stabilized approach—remains elusive. Less elusive if we fly every day or so, but still elusive. Maybe it's a little like learning to ride a bicycle. We just learn.

While this example involves instrument flying, I believe it also indicates that even VFR we can learn proper approach control by osmosis. Bill Lear once had an advertising agent, Norman Warren, at a time when advertising agents weren't above doing a little firsthand selling. Approach couplers were just coming into the market, and Norman had a Bonanza with the L-2 autopilot and coupler. He did not have an instrument rating, but after a couple of years of demonstrating his equipment, when he went up for an instrument rating he needed virtually no training. He'd seen so many precise coupled approaches that he'd learned that perceiving the need for corrections early and making small ones were the secrets of expertise in instrument approaches.

I think in our visual approaches, after we get to know our airplanes, we subconsciously compare and correct during each approach in an effort to make that approach along the flight path which our memory tells us is the correct one. And we even learn to make the needed angular allowances in projected flight-path steepness to compensate for varying wind conditions.

It can be done. In the summer of 1934 there was an air show at the Pontiac, Michigan, airport. I'd had a good summer up there selling C-3 Aeroncas to FBO's, and for the air show, I'd managed to get the factory to supply $100, cash-in-hand as a prize for a C-3 spot-landing contest.

On show day there were eight C-3's on the line, side by side, and they sure looked pretty rocking gracefully in the breeze.

The landings were made from a 180° side approach. Cut the throttle on downwind opposite the spot, no clearing of the engine from there on, no holds barred on sideslipping, or on fishtailing the airplane after round-out. But once you landed, the roll to a stop had to be straight. Whoever stopped with his tail skid beyond but closest to a line across the runway got the money.

In this stage of the economic blizzard of the thirties $100 was a substantial sum of money.

Some of the contestants (the airplanes had no brakes) would hold the stick back and wiggle the rudder rapidly in their roll-out, and some, landing a little shorter than intended, would hold the stick forward just enough to keep the tail skid an inch or so above the runway. It turned out to be a gladiator's contest. I'm glad I wasn't in it. They all stopped beyond the line. The farthest away was just under six feet and the winner was barely an inch beyond the line. I wish I'd thought to ask them how they did it.

THE NIGHT APPROACH

Plotting a proper approach path seems a formidable enough task by day. How do we handle it at night? At night our most commonly used method of staying on the correct approach path angle, at a proper airspeed, derives from the perspective we get from the convergence of the runway boundary lights and their slant in the up-and-down sense. I believe, however, there's a still useful remnant of our daytime sighting angle down.

A correct local or area altimeter setting is a prerequisite, but with that I think we can make up for our lack of ability to judge accurately our height above ground at night. We can still judge distance to the field, as in the day, but at night in many really dark area situations we can be 200 feet above ground a mile out on final instead of twice that and not be so aware of it. With a correct altimeter setting and knowledge of the airport elevation we can come toward the field at pattern altitude until we reach a point where we get the same downward sighting angle at the runway that we use in the day. Which is at least a backup for the convergence and slant of the runway boundary light pattern.

ON FINAL IFR

Our approach-phase experiences taper off with fewer but more frequently serious situations when operating IFR. In an average year there may be around 25 cases in this category. Mostly they involve pilots flying into the ground or colliding with obstacles on final in nonprecision approaches, this is, without glide-slope receiver. Or in other words, in localizer, VOR, or ADF approaches.

At a 2000-foot elevation field, the minimum descent altitude may be 2500 feet. The ceiling is reported to be 500 feet. So we figure we should break out at 2500 feet. At 2500 feet, nothing. We go around. On the next try we go to 2400 feet, get an occasional glimpse of the ground, and decide that a bit lower should do it, so we go still lower, looking out more than at the altimeter.

The 2500-foot MDA on our chart in this approach does not mean that the ground is level from the final approach fix to the airport. In fact it may be an indication of higher terrain or obstacles between the station and the field that reach up within 200 feet of our 2500-foot MDA. When we start shaving the MDA, these things can get in our way.

You might find it hard to believe, but there have been cases of corporate jets flown by high-time pilots flying into wires only 50 or 60 feet above the airport elevation a mile out from the airport on final—when the MDA called for their being 500 feet above airport level over the point of the collision.

I always liked Bob Buck's rule by which he took many and many a DC-3 into Newark Airport with never an incident when they were reporting weather at minimums. Minimums were 200 and a half on the four-course low-frequency range with the station a couple of miles from the end of the runway. On crossing the station and getting the cone of silence followed by A/N reversal, Buck would go immediately down to his MDA of 211 feet and if he wasn't contact up he'd go. No optimistic flying level toward the field to see if they really did have 200 and a half.

Even in full ILS approaches we occasionally get into trouble chasing the glide-slope needle, come porpoising in, lose too much speed, and get into the stall/mush or else collision category. Sometimes the latter is a result of the pilot's getting a glimpse of the ground or lights, which causes him to try to duck under the overcast as in a nonprecision approach.

The point to remember is that none of these things happen to pilots who stick to the numbers.

Finally, there are each year a few cases of VFR pilots who had a smattering of instrument training way back but no recent experience. They often calmly ask for a radar approach, only for it to develop that they can't maintain heading, altitude, or rate of descent within practicable limits. This gets worse the lower they get and often results in a spin or spiral dive.

There are maybe only 10 or 11 a year, but if anything does happen in a go-around IFR, it is likely to be serious because it will probably involve loss of control. It is one of the highest-pressure situations in our flying.

At DH or MDA, we declare a missed approach and, with gear and most likely full flaps down, give it full power and intend to raise the nose enough to be sure we won't fly into the ground. The additional power may cause a pronounced pitch-up moment, depending on how we were trimmed, so we have to hold a lot of forward pressure on the wheel to stop that and avoid a stall. Our right hand may be busy with the power controls or the trim tab, and the tower may call just then wanting to know our intentions. We may also start feeling an urgent need to review the missed approach procedure written on the approach plate, having failed earlier to do this because reported weather had made a missed approach seem most unlikely.

Anxious to get the gear and flaps up so as to be able to really climb, we may get overeager and do it too soon and have to deal with yet another disconcerting trim change, which we have to manhandle until we can get back to the trim tab. Meanwhile, airspeed and heading are all-important.

Unless a pilot is flying instruments regularly and practicing go-arounds under the hood with some regularity, a go-around IFR, as you can see, can be a touchy situation. In go-arounds, I place great store on the Teledyne Instantaneous Vertical Speed Indicator. Its level-flight horizontal up/down needle can be controlled positively with the elevator, and it is free from reverse indications which the standard rate-of-climb indicator has under varying G-loads. By accelerating my scan between the DG and the IVSI, I find it relatively easy to put the IVSI needle on, say, a modest 200-

FPM rate of climb for 30 seconds and do whatever it takes with stick and rudder to hold the approach heading and this rate of climb. With a little breathing space above DH or MDA and in a positive rate of climb with some increase in airspeed from the shallow climb—that's soon enough to start getting the airplane cleaned up, trimmed, and to answer the tower's second or third call.

Otherwise, without an IVSI, full power, heading, nose raised and held only a few degrees above level attitude, or a bit more if necessary to get a first indication of a gain in altitude from the altimeter and rate-of-climb indicator. Hold this attitude firmly until a couple of hundred feet have been gained above MDA or DH in a straight climb on runway or approach heading.

8

LEVEL-OFF
TOUCHDOWN

It's no wonder that in terms of piloting skills, the secret ambi-
tion of every pilot is to excel in making good landings. We
learn early that no matter what anybody says, that's the hard
part of flying. The emphasis in training is more on airwork,
coordination, precision maneuvers, and such. These skills are
important and presumably provide us the ingredients of good
approaches; we know, however, that while a good approach
is the foundation for a good landing, it doesn't necessarily
guarantee a good landing. Knowing good and well that getting one softly onto the ground isn't easy, let's look carefully
for the whys and wherefores of what happens 550 times a year
too often in those last six feet or so down and onto the runway.

DOWN AND LOCKED?

First for the easy part. Of these 550 aircraft, some 75 are
involved in unintentional gear-up landings. The answer to

that is, of course, always to be sure to put the gear down. How do we manage not to? It's not too difficult. There's not been a gear warning system devised so far that can't be circumvented.

For instance, the most sophisticated of all used by Piper, which at approach speed automatically puts the gear down, was also designed to prevent premature gear retraction on takeoff. If a pilot raises the gear switch just as he rotates, or even before, the gear doesn't come up until he's reached maximum rate of climb speed or even a bit more. If the climb gets too steep the gear goes back down automatically even with the gear switch still in the up position.

For short-field situations in which an early-as-feasible gear retraction might be desirable, the automatic system can be disabled by moving a small lever, which latches in the up position. That means after the takeoff, to get back into automatic mode the pilot has to rearm the system. An occasional one forgets to do this and subsequently makes an unintentional gear-up landing. Piper's automatic system is standard equipment on their smaller retractables and has saved many an owner the expense of a gear-up landing. But with the very desirable disabling feature, it's not 100 percent. Beech has at times offered an automatic system for the Bonanza but only as an option, and it has never achieved notable popularity.

Throttle stops have also been tried. I thought the one in the Culver Cadet had all the answers. In the climbout, throttle forward, when the gear was raised, a pin extended out behind the carburetor throttle arm. This would permit retarding the throttle to about 40 percent power, but if you forgot to put the gear down, the stop would be hit and you'd have enough power after flaring to sail merrily across the airport with around 80 MPH indicated (pulling hard on the throttle and wondering why it liked to fly so much).

But wait. Flying a Cadet one day, Howard Piper did a power-off stall, gear down. Then he decided to see how it went power off, gear up, so he raised the gear with the throttle still closed, which put the pin in front of the carburetor throttle arm. That did it. After the stall he couldn't advance the throttle far enough to get more than a fast idle, so he put it down in a random field with only minor damage. Which is quite a feat in the mountains of western Pennsylvania and not to mention on a first flight in a Cadet.

Most of us depend, albeit with a degree of irritation at times, on the old warning lights and gear-horn system. The warning lights are often down around the pilot's ankles when they need to be up at the top of the instrument panel in front of the pilot.

As to the horn, which blows with the gear up and the throttle retarded about three fourths of its travel, what beats that system is putting the flaps down first and forgetting later to put the gear down. This catches twin-engine more than single-engine pilots. The former tend to make slightly undershot, dragged-in approaches anyhow and once they get the flaps down first, it takes so much power to control their approach path and finally drag it in level that they do not get the throttles back to horn-blowing position until they make a last-second throttle-closed touchdown. At which point the horn blows, but, alas, too late.

With no foolproof gadget to do the job, pilots are left to their own devices. Pilots have used various memory systems, such as a red clothespin they move around from throttle to gear lever.

One amphibian pilot had what he felt was a sure bet in the

one type of airplane for which a gear warning system defies invention. He had two full-color slides side by side in small windows in front of him on his instrument panel. With the gear up a light came on behind the beautiful over-the-nose view of an approach to a lake. Gear down and the light came on behind an over-the-nose view of a long runway just ahead. He sent me a picture taken in the cockpit with the over-the-nose lake picture lit up but over the actual nose of the airplane was a runway on which his Goose was resting on its keel, wheels up. So.

BUILT-IN GEAR AWARENESS

Possibly a strongly formed habit can give us as much protection as anything else. For instance, always putting the gear down on downwind after getting slowed to gear-down speed is a good habit. But if something disrupts this sequence which might cause you to raise the gear after putting it down, look out. Such as the tower asking for a long extension of the downwind leg, or requesting a go-around on final approach. Or anything else that disrupts your routine. A door popping open in the climbout has caused many a gear-up landing.

Of course, we don't always go through a downwind phase at controlled fields and are sometimes cleared straight in. Other times we get our attention fixed on vector headings. The need to put down approach flaps early in some of the cleaner airplanes also tends to circumvent the protection of the gear-up warning horn. We need a place that's available in every approach onto which we can hook our gear-down habit.

One such place in an instrument approach occurs on intercepting the glide slope or crossing the final approach fix where putting the gear down gives a usually desired 500-FPM rate of descent.

Habit, however, should not have to be restricted to one place and even only one check for gear down and locked. I was surprised once to learn that a pilot going through a proficiency check for a lower insurance rate at a prestigious advanced training school was criticized severely for making a second gear-down check on final. He was told that putting the gear down and checking on downwind or wherever he preferred to do it each time was enough and that unless he had enough confidence in himself to do the job once and go on to other things, something was wrong.

Every man to his own taste in gear-up landings, but a second gear check on final seems to me to be a capital backup habit. We can't land without going through the final approach stage, so a double-check on final can be at a reliable and unvarying time and place for our habit.

To the 75 unintentional gear-up landings, an additional 25 gear-related cases can be added to cover intentional gear-up landings. These include forced landings, sometimes in water; landings with a partially extended gear or gear down but not locked due to failure in the gear-actuating system; and gear-up landings from failure of the backup system or from lack of any backup system at all for putting the gear down.

Which brings us to looking for some answers to the annual 450 airplanes in which the pilots made it to the airport, but something went wrong in the last few feet and seconds of the descent.

The official statistical tag on nearly all these misadventures is "hard landing." Which suggests that maybe the pilot leveled off too high and wound up "landing" 20 feet or so up in the air. Or that the airplane was flown into the ground in a too slow or too late round-out. But these are minority examples. Such a large percentage of these mishaps occur so far along the runway that they suggest an approach that led to pressure on the pilot to somehow get it slowed and on the ground before running out of runway.

THE PLANNED APPROACH

One of the oldest and truest sayings in flying is that a good approach is the best guarantor of a good landing. A pilot has to supply two basic ingredients for a good approach. First he must pick out and start down an approach path to the near end of the runway the steepness of which is within the control capabilities of the airplane with a little less or a little more power available along the way to stay on that path. Then he has to exercise precise speed control, that is, keep it on 1.3 times stall speed. It is, of course, possible to stay on our selected approach path at a considerably higher speed, but that will mean floating halfway across the airport after round-out.

We tend to come in too fast because we're afraid of stalling, but we shouldn't be. At 1.3 times stall speed we are beyond reach of atmospheric effects on angle of attack in slower flight, yet not too fast to land in the first third of the runway if we've stayed on our selected flight path. Excess speed in an approach serves no safety purpose. At a proper approach speed, with partial power or power off, the nose is going to be 4° to 8° below level attitude, and no airplane is going to stall in a stabilized approach in that attitude.

There is more than one way to land an airplane and those ways can be multiplied by the two kinds of airplanes we fly: those with the third wheel under the nose and the mains back behind the CG of the airplane, and those with the third wheel hooked to the tail of the airplane and the mains ahead of the CG. The tricycles and the taildraggers. In a roll-out the tricycles want to roll straight because the CG is ahead of, and you might say, pulling the main wheels. In contrast, the taildraggers in a roll-out, with the main wheels ahead of the CG, want to turn and go tail first, putting the CG ahead of the mains, à la tricycle. The tricycle tends to be stable rolling and the taildragger unstable, until it does its 180° dido.

How to handle them: let's look at some of the different practices in landing airplanes, not in terms of what is best but how they do or at times don't work.

In any repetitive activity human beings have a talent for finding ways to reduce the amount of effort required. I believe this accounts for the most common type of landing you see today, especially on airports with any appreciable amount of training activity.

The airplane, a tricycle, comes in low and fast, completes its round-out only a few feet above the runway, and the pilot cuts the power and simply holds the airplane in a level attitude from there on, waiting for it to settle onto the runway, nosewheel and mains simultaneously.

This type of level-off touchdown is not only economical of work but also of skill. Unfortunately it is quite a reportable-accident producer. While it is possible to get many soft touchdowns in this manner, it is possible to get hard landings from leveling off too high and dropping in too hard. The procedure also requires a lot of extra runway because the approach speed needs to be 10 to 15 knots over a normal approach speed in order to stop the sink in the flare with little in the way of a nose-up attitude at the end of the round-out. Obviously the procedure doesn't work well on a short field because the touchdown is so far along the runway, and at way above tail-low touchdown speed.

THE BENT NOSE GEAR SCENARIO

The most common misadventure to result from using high-speed level-attitude touchdowns is the nosewheel-first landing. Occasionally a pilot fails to complete his round-out and overloads the nose gear. But more often, and more often than you might think, it is just a case of a pilot's getting tired of floating merrily along the runway and deciding to put the nose down and fly it on. The nose gears on tricycle airplanes aren't designed to withstand the high-impact loads this procedure can generate.

Over a not-too-long period I've seen this happen several times. First there was an elegant Bonanza that floated halfway along the lengthy runway at Lake Tahoe and then, from about 10 feet above the runway, it was nosed down. The nosewheel folded back and the prop blades curled. And there was a Skylane on a windy day landing on a long runway at Lakeland, Florida. Same thing. More recently a student in a Cessna 152 landed nosewheel-first halfway along a 6000-foot runway. The nose pitched up, followed by a more nose-down landing and pitch-up. The third landing was still more nose down. The nose gear folded back and the airplane stopped nose down, tail high. This type of landing is not, however, primarily an affliction of student pilots. While high-speed level landings appear to be the trend of the times in pilot training, often pilots who were taught to land tail low no matter what they were flying and who have been flying a long time follow the line of least resistance and put one on nosewheel first and too hard if they start running out of runway.

The revival of the tricycle gear was supposed to open an era of easier and better landings, freeing us of the more demanding judgments required in making a three-point, stick-full-back touchdown in a taildragger. Somewhere along the line in our educational system we failed to explain properly the tricycle gear's landing characteristics. The tricycles are easier to land because, while they need to be landed tail low, in contrast to the taildragger, there's no critical value on how low. The lower the better in many situations, but the key thing is to get the tail low enough to avoid a nosewheel-first touchdown. We forgot to stress that. And it's costing the owner dearly.

LANDING FAST AND LEVEL

A variation of the high-speed, level-attitude touchdown style of landing is completing the flare to level flight a bit higher, say 10 to 15 feet above the runway, leaving the power on a bit, and then landing with the throttle. That is, by holding the attitude level and keeping the rate of sink low but constant with minor power variations as needed. The FAA seems to have a predilection for a pilot's flying the engine rather than the wing at this juncture. Landing with the throttle in this manner can get even more soft landings than the lower-flare, level-attitude, power-off touchdown. But it sure eats up a lot of runway.

The high-speed level landing has some positive factors. For instance it mellows the gust effects, which can produce extra lift momentarily, or reduce it. But in its call for a lot of runway it invites heavy braking. It's seldom possible to rent an airplane used in training, or even get a cross-country rental, without getting a shake right after lift-off from unbalanced tires resulting from flat spots on the mains caused by too-heavy brake application.

How strong would nose gears have to be to withstand higher nosewheel-first landing impact, since tricycles seem to tempt pilots not to get the tail down in a hold-off after the flare? When the Pipers put a nose gear on their doughty Pacer and christened it Tri-Pacer, the tricycle gear was just hitting the "mass" market. At Lock Haven they started getting complaints that the nose gear was not strong enough. Pilots were landing nosewheel first and bending them backward.

Cessna had also gone the tricycle route with their 170, evolving it into the 172, but they weren't having as many nose gear experiences as Piper. Howard Piper, surely as consumer-oriented a test pilot as there ever was, rented a 172 to see how much stronger its nose gear was than the one on the Tri-Pacer.

He'd been banging the Tri-Pacer on nosewheel-first pretty hard without a failure. In his first nosewheel-first landing in the 172, in what he considered to be a not unduly hard one as compared to what his Tri-Pacer would stand, the 172 came to a stop with the nose gear "retracted." For which he had to pay. Nose gear strength obviously wasn't the only factor. In hindsight it seems that Tri-Pacers tended to inhabit more small fields than the 172's. They also had only moderate flaps and a higher approach speed than the 172's. So on the high-speed level landings, often on too short fields, there was more frequently pressure on the Tri-Pacer pilot to get it on the ground. All of which is an example of things manufacturers can learn only by experience, since it's the customer and not the test pilot who puts the hours and years on the airplanes.

ANOTHER APPROACH TO LANDING

Now let's look at just the opposite of the "fly it on fast and level" (and sometimes nosewheel first) philosophy, namely, the no-flare landing. Or maybe there is a slight flare just before touchdown. But the final approach is made in an essentially tail-low touchdown attitude, with just enough power to keep it hanging on the edge of a stall and to keep the rate of descent within the shock-absorbing capability of the gear. Landing gears, that is, the mains, are designed to absorb within their travel limits the impact loads in a several-hundred-feet-a-minute touchdown with little or no rebound. My interest in this type of landing is wholly academic but I think it worth knowing about for perspective.

On the front door of Cessna's Pawnee plant in Wichita is the most heartening sign any salesman will ever see: "Salesmen Are Welcome." Inside the lobby is an attractive showcase display of *Detroit News* trophies. They were awarded annually back in the mid-thirties following a contest for the world's most efficient airplane. Each year the bacon had been brought home by Dwane Wallace flying his dream airplane, the 145-HP Warner-powered four-place cantilever-wing conventional-gear Cessna Airmaster (145-MPH cruise at optimum altitude). With this airplane he was beginning to get the company out of the economic overcenter situation which had brought it to a stop soon after the financial blizzard which began in 1929.

One event in the *News* contest, which resulted in some spread landing gears and disqualifications, was to see who could approach over a 50-foot barrier, land, and come to a stop the shortest distance beyond the barrier. Dwane always won, stopping just under 1000 feet from the barrier. In, mind you, what would be considered even today a fairly hot-landing airplane.

Only recently did he tell me how he did it. From 400 feet on down in the approach, they required that the throttle remain closed. On downwind he'd pivot around the nearest barrier post (there was a ribbon between the posts), put his split flaps down (they created drag but little lift) and finally aim just above the barrier in what he estimated to be a 400 to 500 FPM rate of descent at a relatively low airspeed. He had no rate-of-climb indicator. Going into the ground cushion without an attitude change helped reduce the rate of sink, but naturally he had to have enough speed left for a last-minute flare into a full three-point attitude. Crunch, with no rebound. And immediate heavy braking. In short, he crossed the barrier in a steep descent with the wing on the edge of a stall but with enough flare capability left to reduce the rate of descent to within his design limits of the gear. All of which I can understand, except how he could do it without power and never a miss.

My own first look at this type of approach and touchdown was in 1954 at the Army's Fort Rucker in Alabama. Helicopters were everywhere, also Cessna L-19's, or fixed-pitch-prop tandem-seating versions of the Cessna 170 with a bigger engine.

THE STOL APPROACH

The L-19 training area was mainly a lot of small clearings surrounded by trees, and they all were tight fits for a fixed-wing aircraft. The approach? Nose well up, power considerable, approach path steeply down. What shook me was being told that anytime a cadet lowered the nose in one of these approaches, he got washed out. Attitude fixed, constant. Power could be used at will to control rate of descent and consequently glide path. One after another they'd hit hard, three-point, sometimes with a lot of power on at the last, and stop short.

I don't think this is any way for a general-aviation pilot to fly, because in such an approach the airplane (and pilot) is too much at the mercy of the vagaries of the wind. Resulting eventually in a too hard landing. Behind the maintenance hangar several L-19's rested on their bellies, spring-steel gear spread out horizontally. On the bottom of each wing there were black tire marks—imagine, on the bottom of a high wing! Obviously the pilot had been unable to stop a too high rate of descent with power.

My first actual experience with this power-on, no-flare or virtually no-flare tail-low three-point touchdown was in a Helio Courier designed by Otto Koppen. The purpose of the design was not only to achieve the usefulness of a steep take-off and slow landing vehicle. Professer Koppen wanted also to make flying barely above stall-speed safer. Light wing-loading, big slow-turning prop, full-span leading-edge slots that would automatically extend forward when the airspeed got down to around 55 MPH, long-span high-lift flaps, short-span wide-chord ailerons with lots of differential up-travel, and long-travel oleos in the taildragger gear mains.

I rode on a demonstration flight in a Helio Courier in the early fifties at Flushing Airport, near New York. The pilot was Lynn Bolinger, a Helio investor and Harvard Business School professor who almost convinced the dean that general aviation was about to enter the realm of legitimate business and had great promise. Lynn could really fly the Helio.

On final Lynn would put the prop in flat pitch and set the throttle to get 1700 RPM. Then he'd lower the flaps and raise the nose gradually until the slots popped forward with a thump. Then he'd set the nose a bit higher. The approach path was so steep it was obvious about where the touchdown point would have to be. In this flight regime the airplane still had quite responsive roll, yaw, and pitch control. With any undesired increase in sink rate Lynn would simply add a bit of power, or reduce power to increase the sink rate, maintaining meanwhile a constant attitude.

The touchdown, three-point, still with 1700 or more RPM, was firm each time, but not really hard due to those long-travel oleo gear cylinders. Immediately on touchdown, heavy braking would raise the tail to level and we'd come to a stop with the tail sinking gently back to the cinder runway—only two fuselage lengths from the initial touchdown point. Into no more than a 10 MPH wind.

Lynn told me later that he'd never encountered a down-draft or gradient-wind effect he couldn't offset with an increase in power, what with the high RPM starting point he used.

The Helio was a remarkable machine, with a limited special-purpose market.

But, like the L-19's, it, too, could get bruised by a pilot who was too slow in getting the power up in response to an objectional increase in rate of descent, or if there were enough of a gradient wind or expiring gust at the last that the pilot didn't have enough reaction time. I've seen Helios in the shop with the mains and tail wheel on the floor as well as the middle of the fuselage—which indicated a hard enough landing to break the airplane's back, as it were. There were also incidents in which the airplane would be overdemonstrated in takeoffs started at midfield on short fields. Rotated into a steep climb attitude after a short run, at the top of the ground cushion, nose still way up, it would start moving horizontally forward and fly, on occasion, into treetops.

The Robertson conversions are of the Helio genre. While they also avoid stall/spin incidents, I've never found an owner of one of their conversions who wouldn't admit to an occasional unexpectedly hard landing. The STOL's can't escape entirely the vagaries of the winds.

There were, of course, the Grasshoppers initiated by the Piper Cubs, which, while not truly STOL aircraft, were flown as if they were in low-speed, power-on approaches with power-on three-point touchdowns after a last-foot flare. One day at Flushing Airport I learned something about how at least one pilot judged angle of attack in such an approach. Thomas Piper was conducting a press demonstration of a new model Grasshopper J-3, which had enormously oversize balloon tires in tandem on each main landing gear. He'd come in steeply, touch down in the tall marsh grass maybe 50 feet short of the runway, burst out of the bullrushes, and stop in the first few feet of the runway.

Someone asked Tony how he judged his speed in such an approach and landing. His words were along the line that you have to have something in the bank. What he meant by that was that he used how far the stick was from all the way back as an indicator of adequate speed for a flare. If the stick was too far back before the last-second flare was started, you'd find there wasn't enough elevator left to flare. Which wasn't so good with a shock-cord gear.

THE AIRLINES' APPROACH

To complete the picture of the nether end of landing tricycle-gear airplanes with little or no flare, every time I see an airline jet land, I try to detect a last-second flare just before touchdown. It seldom seems to be there. Some airline pilots say they do flare, but most admit that they do not. With everything they have hanging out on the wing in an approach, they are flying the wing at a high angle of attack even though the fuselage is in only a barely nose-up attitude. With the proper approach speed for the forthcoming landing weight, which is calculated each time, they simply hold that speed and their shallow deck angle and go into the airplane's outsize ground cushion with as much, in the 747, as 700 FPM rate of descent at 140 knots. They depend on the compression effects of the wing in the ground cushion to increase momentarily the density altitude and thus increase their lift and slow their rate of descent to an acceptable level prior to touchdown. Some airline pilots mention that in a last-minute flare, what with the main gears so far behind the lateral axis of rotation, the wheels would be moving downward faster than the airplane. Instead of flaring at times just before impact, they ease the nose down a bit so the wheels will have a lower sink rate than the airplane when they hit the runway.

One captain told me that he had at times on ferry flights held the airplane off after the flare and landed it like anything else and actually would prefer to land that way all the time. But he didn't dare because if anything happened he'd be in trouble if the flight recorder showed he wasn't doing it exactly by the book. Still another captain said that he preferred not to flare because of the abrupt rise in drag when a swept wing is tilted into the higher angle-of-attack range. Which might give a real drop-in.

At any rate, the big jets do hit hard much of the time, or at least hard if you're sitting over the wheels rather than in the nose. On one of my few domestic airline flights, to Alaska a year or so ago, I didn't see a soft landing anywhere along the line. They were all very much on the hard side, some alarmingly so. Which tells me that the big ones, while maybe less so with their higher approach speeds, are still subject to the atmospheric ups and downs that our small airplanes experience in exaggerated form.

Once in a while someone gets into the rule-making circuit

who has never flown anything other than a jet and figures that everything should be flown the way a jet is flown. When the person is a rule maker, it can get the rest of us doing some wrong thinking. That doesn't allow for the variables the general-aviation pilot has to deal with in flying a machine with light wing and high power loading.

Speaking of no-flare landings, which, of course, would have to be in a taildragger to be successful, one night in the early forties a 20-foot-thick layer of dense fog lay on the airport at Indianapolis. A TWA pilot, flying a DC-3, flared as he descended into the layer and damaged the airplane in a hard landing. New procedure notice: in a situation of this kind a no-flare landing should be made. Bob Buck, then Chief Pilot at TWA, started checking their pilots out at night at La Guardia on this.

I had the opportunity of watching one night, standing in the aisle between and slightly behind the pilots. It was most interesting. The procedure was to set up a 200-FPM descent and hold it right onto the runway. Immediately on touchdown, move the wheel forward about an inch, hold the runway heading, and let the tail come down by itself as speed was lost. At first, at the usual height the pilots would start to flare, and Bob would keep repeating, "Don't do it. Don't do it." By the third approach the pilot wouldn't and from there on was in business. The landings weren't too hard but I did find that keeping my knees flexed made them softer and gave me a feeling of participation.

Next day I called Wolfgang Langewiesche and told him I had seen a new way to land. He's easy to catch if the trap is baited with flying. In an hour we were at Flushing Airport and had rented a new Aeronca Chief from Speed Hanzlik. Pretty soon Speed was climbing the wall, muttering, "What are those spooks trying to do with my airplane!" We were making TWA-type no-flare landings, and they worked just as well as in the DC-3. In fog, calm, stable air can be expected, and fortunately we had only a very light wind. Taildragger pilots, of course, make high-speed level touchdowns, usually in high wind conditions, but they flare and keep it level in the roll-out as long as they can with gradually increasing forward stick. The difference in the fog-layer procedure described is that there is no flare with a quite low rate of descent.

A friend asked recently that I make a trip with him in his Citation and rate his crew's operational practices. The captain had been a bomber pilot in World War II and had subsequently gone into corporate flying and eventually turboprops. The copilot, young and lucky, had first acquired a mechanic's license, airplane and engine, and subsequently his commercial license, multiengine, and instrument ratings. He takes care of minor maintenance and a continuous and thorough inspection routine. Both of them had been through the Citation school at Fort Worth and had been flying the Citation almost a year, around 75 hours a month. As far as I could tell, their procedures were on a par with those of the trunkline airline pilots, who in 1980 made a takeoff every six seconds and flew 225 billion seat miles without scratching a passenger.

The Citation crew had a small box down at one corner of the flight log form, which had no legend. These boxes would have a 1, 2, or 3 in them, which aroused my curiosity. On nearly each flight they alternated as to who would be on the left side, and the last entry on the form was how the pilot in the right seat rated the landing. One, thumbs up. Two, thumbs maybe only half up. Three, thumbs down.

Which caused me to ask the captain why they preferred landing each time with the tail only slightly down. He said that in holding for a tail-well-down landing, the airplane eats up too much runway and that they preferred to get it on the ground earlier so that reverse thrust could be used. He felt this always gave them an earlier turn off the active.

THE MIDDLE METHOD

Between the high-speed level landing and the no-flare landing a new constant-attitude approach procedure seems to be becoming popular, a sort of middle-of-the-road proposition. The constant attitude is used simply for the smoother ride on final but not to eliminate a normal flare and tail-low touchdown. At the starting-down place the nose is put a few degrees below level attitude and the power is reduced to not much more than a fast idle, or, say, 12 inches MP with a constant-speed prop. From there on the attitude is not allowed to change. Power is varied as needed to keep the airport in reach. Increased power flattens the glide path with the attitude held constant and in moderate amounts gives a surprisingly low increase in airspeed. I find the procedure rather convincing except, still new at it, I occasionally wind up with the attitude unchanged, the throttle closed, and not getting down. Or, in other words, in a too high power-off approach. The cure for that would be to move a little more toward the Army's Fort Rucker procedure and use a bit slower approach speed and more power.

All of these approach and touchdown variations leave those of us who prefer to try to hold a constant airspeed on final with the elevator, and use the throttle for altitude control, giving our passengers at times a roller coaster ride in turbulence. But we have the advantage of keeping our angle of attack a constant 6° to 8° lower than stalling. And we do our share of leveling off too high and dropping it in too hard—

or, in taildraggers, dropping it in and getting a pitch-up because the stick isn't full back, followed by an all-thumbs recovery from the bounce. More abandoned lousy approaches could lead to a lot fewer hard landings.

You will have observed that all of the foregoing approach and touchdown procedures, while possibly best for some airplanes in some situations or for some specific purpose, have shortcomings. The figures show that whatever our system, we have a surprising lot of trouble in those last few feet of our descent.

Whatever our preferred procedure in the level-off touchdown phase, we need to have in mind its strongest weakness. For instance, in the high-speed, level attitude landing the touchdown point and runway length are critical factors. Getting a slightly nose-up attitude after the flare and from there on "landing" with the throttle also calls for extra runway length.

My own possibly somewhat out-of-date preference is for tail-low touchdowns and for precise speed control in the approach. The latter admittedly can give a roller-coaster ride sometimes, but in our general-aviation aircraft I think that is preferable to either a high-speed approach or fixed-attitude approach. Flying into an atmospheric sinkhole on final I'd rather be with the fellow who not only guns it but also noses down a bit before getting back on his selected descent path. When it comes to touchdowns, at least on fields of 3000 feet or less, or on the short side for any given airplane, if the airplane isn't on the ground in the first third of the runway, it's time to go around. But in the end, isn't the need of a go-around this late an indication that not enough emphasis in our training is put on abandoning a poor approach before rather than after the flare? This should instill better judgment than trying to get one on willy-nilly after coming in too high or too fast or both.

9

THE LANDING ROLL

Now we come to the last phase of our takeoff-and-landings hegira. Following a presumably uneventful touchdown, some 550 airplanes are substantially damaged each year in the subsequent landing roll. This classification does not include any ill-starred go-arounds. Those cases are back in the initial-climb category discussed earlier.

What happens? Quite a few land long and go into the fence or bushes or ditches at the end of the runway, or do an intentional swerve or ground loop, trying to duck the punch. Some land in snow and go out of control, getting a main into a low snowbank often being a factor. Some land in soft ground and get out of control; of these many are taildraggers which nose over when, with the tail starting to come up, the pilots increase brake pressure instead of getting off the brakes, and, stick full back, add a bit of power to blow the tail down. Then there are always a few cases of pilots making intentional off-airport landings on too rough ground, or on beaches, and losing control. All of this, however, concerns small numbers.

CATCH YOUR DRIFT

The big number in the landing-roll category bears the secondary tag "ground/water loop/swerve." You can forget "water loop" because there's so little seaplane flying, so for landplanes the designation should be only "Ground-loop/-swerve." And since taildraggers, which are in the minority, ground-loop but tricycles do not, then "swerve" is what applies to most of our landing-roll difficulties.

It is my conviction that our ground-loop experiences in taildraggers and swerve experiences in tricycle-gear airplanes begin with a touchdown with drift. The effective techniques of stopping a ground loop early in taildraggers and using the nosewheel and centrifugal force to pick up a wing in a tricycle if it tips on touchdown by turning the nosewheel toward the low wing have been examined earlier. So isn't what we're really looking for an improvement in ability to touch down without drift? Whereupon I risk excommunication.

In order to touch down with no drift, it is, of course, necessary first to recognize the drift. It takes a while before a person who is learning to fly begins to see drift. Thing is, it doesn't take much drift on touchdown to upset the applecart, and the smaller the amount, the more difficult it is to catch in our peripheral vision.

It's a wonder new pilots don't contribute more than they do, because pilots mainly have to teach themselves how to handle crosswind landings. A part of that stems from the sheltered life many students lead: if it's very windy, or there's much crosswind, operations are shut down at many schools. Certainly so for student solo flying. How many landings with enough crosswind to really count does the average pilot have when he receives his first license? Not many. And most of those in only moderate crosswinds—just enough to show the pilot understands the principle. But crosswind landing isn't something that will stay swept under the rug.

As you know, the most prevalent system of crosswind landing taught today is the wing-down, sideslip method of offsetting drift—and landing on one wheel, wing still down. The method is disquieting to passengers and in some aircraft the pilot comes sideslipping in on final with a no-sideslipping placard somewhere in the cockpit. To me the method is very unrealistic. If it works for a pilot, well and good, but I can see many reasons why it shouldn't work satisfactorily in other than light crosswinds. Basically, the maneuver looks good from the point of pilot skill, but it doesn't deal with the realities adequately.

A first question is how much crosswind component can a sideslip offset? Say there's a direct crosswind of 15 MPH. Can the airplane be made to drift sidewise 15 MPH in a slip? And how far down should the wing be to do this? If it's too far down, too much of the sidewise velocity will be toward the ground instead of into the crosswind. Seems to me that 5 MPH sidewise movement in a sideslip would be a lot. Item, and a seemingly logical one: on a demonstration flight a couple of years ago with the wind from the left, the prospect in the left seat climbed out carrying the left wing down. Curious, I asked him why he was doing this. His answer was, "To offset drift."

A further concern about a touchdown in a sideslip is how much of a side load a main gear and tire can be expected to take if the airplane is dropped in hard from a too high level-off, or is simply sideslipped into the ground. In drop test for landing gears, the 14-inch drop is onto a flat surface (wings off the airplane and sandbags to get gross weight). This is equivalent to a drop-in from about 11 feet in an actual landing. In this situation the mains divide the load equally. It's just hard for me to believe that either of the mains alone could be expected to withstand the same impact in a wing-down touch-down. Not to mention that in some of the short-legged low wings a wing tip might get to the ground first.

Finally, how well does the wing-down system work in a crosswind landing when the crosswind involves strong gusts? Which asks the question of how did the system originate in the first place. I had been flying 15 years before I'd ever heard of it, and even then it was only in limited use. I don't know where there's any history to read on the subject, but I believe the system had its origin in an alibi.

GROUND LOOP: TAILDRAGGER'S LAMENT

In the 100-percent taildragger era, especially before steera-ble tail wheels, the gears were narrow. They were also well forward to reduce the nose-over tendency when rolling into soft ground. This forward location of the mains made the airplanes prone to ground loops. The slightest side load on the gear on touchdown or in the roll-out could trigger the start of a ground loop. After only a few degrees into the loop it would accelerate abruptly and often beyond control. There were lots of ground loops.

Once it gets going, a ground loop in an airplane with a narrow gear can put a wing tip on the ground sooner than would be the case with a wider gear. I believe this prone-to-ground-loop/narrow-gear combination caused many pilots to really believe that their trouble was the upwind wing having enough dihedral to let the wind get under it. Time and again after a ground loop that got a wing tip on the ground early pilots would alibi, "The wind got under my wing."

I don't think the initial trouble was the wind getting under the wing. Airplanes have always had enough roll rate, even at low speeds, to give full lateral control right to the end of level-off and to the instant of touchdown. The trouble pilots were having was a failure to recognize a ground loop's start early enough and to get on the ball with rudder, power, and brake if the airplane had brakes. It's my conviction that the father of the wing-low crosswind landing system was the ground loop.

At any rate the wind-under-the-wing alibi served to develop a belief that it was a good precaution after flaring to keep the nose straight but the wing into the crosswind slightly below level so that the wind couldn't get under it. They were talking, of course, not really about wind but gust effects. Nothing was said about the wind, or a gust, pushing a down wing farther down if it couldn't get under it. Finally someone must have discovered that at least in light crosswinds, the flare and touchdown with a wing held down could result in a soft one-wheel landing with no drift. The concept was soon embellished by the discovery that the amount of sideslip it took coming on final to keep lined up with the runway was a good starting indication of how much was needed to offset drift.

But to me, on a wild and windy day or a gusty one, the sideslip crosswind landing concept is of limited application. A wing knows nothing about wind other than a change in velocity or direction of the wind it flies into—and that only momentarily, until the airplane has time to accelerate or decelerate and return to its trim speed in the new ambient condition. When it comes to a strong crosswind in a landing, and most especially if there are strong gusts, it seems to me that the crab method offers more and better options.

On final we fly a crab heading as necessary, wings level laterally, to keep lined up with the runway center line. In the last 50 feet or so down the amount of crab needed is likely to decrease somewhat. Holding whatever crab it takes to stay over the center line, we round out holding the crab heading, hold off, float into a low rate of sink, and when that running-out-of-lift feel of an impending touchdown develops, we give it rudder toward the runway heading and opposite aileron as necessary to keep the wings level, centering these controls when we reach runway heading.

Are we then at the mercy of the crosswind? No. There is a golden moment in the procedure at this point. There is no drift. It's a relatively long moment before any drift sets in, and then only gradually at first. In the meantime we have usually touched down if our timing has been right. The pause before drift sets in occurs because, as far as the ground is concerned, the airplane has to have time to accelerate to the full drift component of the crosswind. It can't accelerate to this instantly.

If we don't touch down, preferably with stick back, soon after straightening out to runway heading, and drift sets in, we need to yaw the airplane, keeping the wings level, till the nose is pointing in the direction in which the airplane is moving, which will give a touchdown with little or no drift.

The key element in the crab method is a precise feel for the airplane. After the flare, with the tail well down, we need to know when the airplane is about ready to land. At that point we yaw to runway heading with crossed controls. Next we need acute perception of the start of any movement in drift, so that we can be sure of having the airplane pointed where it is tracking at the moment of touchdown.

Today you see FAA mandated placards in cockpits naming

the demonstrated maximum crosswind-landing capability of the airplane. As I understand it, this doesn't necessarily mean the airplane cannot be landed in a crosswind in excess of the specified velocity but simply that on the day the government man arrived at the manufacturer's, this was all the crosswind available in which to demonstrate crosswind landings. Obviously an inspector couldn't drop anchor waiting for more wind. So what he saw was put on the placard.

For many years it did not occur to me that there should be any limit on how much crosswind could be handled in a landing. Certainly there'd always be a heading that would give the ground track desired. At least I never encountered a crosswind condition that the crab method couldn't handle. But this was in an era of airplanes with small vertical fins and large rudders with lots of rudder travel.

Today our airplanes have comparably large and larger vertical fins, plus an occasional ventral fin, and smaller rudders with less rudder travel. All in quest of more directional stability and less likelihood of a roll-off in a stall. The amount of rudder travel has been reduced because with just a bit too much it is easy to get into spin entry problems. Some of our airplanes also have stretched, flat-sided fuselages, which add to effective vertical fin area as well as weathercocking tendencies from side gusts. If the figure is reasonably high, maybe the demonstrated crosswind placards aren't beside the point. A recent experience in an early 40° flap-travel Cessna 150 made me think so.

There was a variable crosswind on the single runway, from the left, gusting to about 15. The hangar was just to the left of the approach end of the runway being used. I flared in front of the hangar with only a little in the way of a crab heading and floated a bit. When I flew beyond the hangar, all of a sudden, even with full right rudder as fast as I could put it in and opposite aileron to keep the wings level, the airplane simply pivoted about 25° to the left and hesitated there for an interval, still with full right rudder. Only when I gunned it—what else was there left to do?—was it possible to get it back on the runway heading, wings level, in time for a no-drift touchdown.

I think the lesson that day was that with some of today's efficient high-lift flaps, some of our airplanes will fly so slowly that the rudder effectiveness gets below counterpunch power needed on encountering a direct crosswind gust to 15. It never happened to me in some recent happy flying in a Cherokee Six-300, but I've heard the Six is in this category. So times have changed, or at least the airplanes, but I'd still rather rely on the crab rather than the wing-down system in gusty crosswind landings.

THE GROUND SWERVE PROBLEM—SOLVED

There is a curious thing about the efforts that have been made to find mechanical solutions for flying-technique problems. The list of successive failures is a long one. To the list I must add the crosswind landing gear, conceived by John Geisse of the FAA of yesteryear and put into production by Goodyear in the fifties. I flew it on a Cessna, 180LC, for a year. It simply eliminated the crosswind landing problem. In a crosswind, holding whatever heading in the approach and after the flare it took to stay lined up or over the runway, you simply held off and made your tail-low stick-back touchdown with the airplane way off the runway heading. On touchdown the mains would simply caster as necessary to relieve the side load on them. The wheels would point straight down the runway and you'd roll merrily along sidewise, much to the consternation of onlookers, steering with the rudder pedals just as you would with a standard gear, but still rolling straight ahead sidewise, if you follow me. At the end of the landing roll the wheels could be made to return to their straight-ahead detent with a slight turn in the direction in which they had castered on touchdown and in the roll-out and you'd taxi normally, unless you made a too fast turn and the wheels castered again. You could get them back in detent position or simply taxi on in sidewise.

In a crosswind takeoff, if the sideload on the mains was enough to get the wheels out of detent position, you'd simply swing the nose into the crosswind a bit and make a straight but sidewise run along the runway, steering with the rudder.

After a few takeoffs and landings with the gear, you simply quit paying attention to the direction and velocity of crosswinds.

The crosswind gear ate up tires rapidly with a lot of sidewise taxiing but it was fun clowning with it and with precise control, although it didn't look like it from outside. Trying to move the airplane backward sometime gave hangar personnel fits, for the wheels would go slewfoot, and the airplane had to be moved forward to get them back in the detent position. There was one no-no about taxiing along the downwind edge of a runway in a crosswind. In this situation, with enough crosswind to make the wheels caster, the airplane could go sidewise off the runway and the only way to get it back on would call for swinging the tail out into the bushes, or the runway boundary lights. It was also important to know that when the wheels castered to their stops, maybe 20° or 25° either way, the gear then became conventional and the airplane could ground-loop unless the normal directional control procedures were used.

These things and also the several-hundred-dollar cost of the gear might have been what made it unpopular, but to me the joy of being able to land with the airplane on a crab heading 20° to 25° off runway heading, and to roll straight sidewise with no concern whatever, was worth the cost. But, into limbo went a brilliant solution to our ground-loop/-swerve problems, at least in taildraggers.

I asked the people at Goodyear once when they were going to offer the gear for tricycle airplanes. Apparently they'd moved in that direction, but the only explanation I got was that you just don't put a regular crosswind gear on a tricycle and have it work acceptably. Presumably one could be made to work all right, but I doubt by then they felt there was enough of a market for it.

I read somewhere that on an exam at Annapolis there was once this question: You've lost power and are adrift toward a rocky shore. The sea anchor has been put out, but you're still drifting toward shore. What else can you do? The midshipman who answered, "What the hell else can you do?" did not flunk. In crosswind conditions so extreme that a conventional approach and touchdown might seem to be inviting disaster, and with no place else to go, there's something else you can do. And I think it should be preserved for posterity.

THROW THAT CROSSWIND A CURVE

I should have introduced Franklin T. Kurt, one of my several gifted mentors, earlier. If private flying ever had a friend, it was Hank Kurt.

Hank, an MIT graduate in aeronautical engineering, became a naval aviator, and later entered general aviation, operating at New Haven, Connecticut: charter, instruction, and the design and limited production of a sleek two-place biplane, the Kitty Hawk.

From New Haven, Hank went to Grumman. Wherever an

extra hand was urgently needed in a hurry, that's where you'd likely find him: in test flying, R and D, in quality control, or out selling a Goose or Widgeon. He even taught Mr. Grumman's daughter to fly Mr. Grumman's Stinson Voyager.

Now for the Franklin T. Kurt curved takeoff-run and landing-roll crosswind system. His background was given to avoid the possibility of your thinking that it might be a wild idea of some way-out guy.

Here's what Hank would do in anything he flew if there was a strong crosswind, and even on runways of less than average width. For takeoff he'd start on the downwind edge of the runway, get the airplane headed at a point maybe 100 yards ahead along the upwind side of the runway, and pour on the coal. He would then immediately start curving the run toward the center line of the runway, decreasing the curve rate as speed was gained. By the time he passed his initial aiming point, he'd be on runway takeoff heading and would keep curving, but now at an even slower rate. The curved run took all the side loads off the wheels and eliminated any tipping effect. He was using the centrifugal force of the turn to keep the airplane comfortably vertical. He was keeping the crosswind and centrifugal forces in balance, and when it came time to rotate, he wouldn't be too much off the runway heading toward the downwind side.

My impression was, and it proved correct in practice later, that this takeoff run was shorter than it would have been using full rudder and possibly some brake as necessary to make a straight crosswind takeoff. Also, in a strong crosswind the curved run prevents the airplane from starting to drift even with the mains still on the ground as the wing begins to pick up its load, as sometimes happens in a straight-ahead crosswind takeoff. The procedure should not be used with less than half tanks of fuel because of the possibility of a fuel-tank outlet becoming unported. Especially so with rubber fuel cells, which have relatively high unusable fuel restrictions because it's not feasible to use bulkheads and sometimes flap valves in these tanks to keep fuel over the outlet ports in odd attitudes or in a curved takeoff run.

THE CURVED LANDING ROLL

Even more convincing than Hank's crosswind takeoff was his crosswind landing. On final he'd parallel the runway center line maybe 200 feet on the downwind side. Nearing the airport boundary, he'd take a cut toward the runway of anywhere from 10° to 20°, depending on runway width and the velocity of the crosswind. At this point he'd be on final for a no-drift touchdown and would use the crab method or a modest combination of that and the wing-down method for keeping the drift at zero. Just before crossing the downwind edge of the runway he'd flare, hold off, and make a stick-back touchdown and then start his curve out of the wind to runway heading, slowly at first, then faster as speed decreased, in order to keep the crosswind and centrifugal forces of the curve in balance. With a strong crosswind component he'd continue the curve past runway heading 10° or so.

Don't wait until you're in extremis to try this system. Explore it first in moderate crosswinds with their associated lower initial angles and rates of curvature of the run and roll. It is obviously not a good system for the high-speed, level-landing pilot, as it calls for tail-low touchdowns without excessive floating.

Admittedly the procedure's description might invite visions of running off the runway at the start of the takeoff or shortly after a touchdown off runway heading, but you'll find this doesn't happen. I've used the system in high crosswind conditions time and again in Cessna 180's, Skylanes, 210's, and Comanches, Twin Comanches, and Aztecs without any problem about curving the run or roll fast enough to avoid running off the upwind or downwind side of the runway.

Practicing the takeoff procedure a few times before trying the landing one will make the latter much easier.

As a secondhand dealer in the obvious let me add that, with today's wide gears, including the remarkable tapered tubular steel ones on the high-wing Cessnas, the primary virtue of the Kurt system is what those cuts into the crosswind on landing final do to eliminate or greatly reduce the need to take out the drift. For instance, in a crosswind that is quartering to the runway, the cut on final may eliminate any need for drift control, or if not that, reduce it to a minimum. In a direct crosswind to the runway, a 10° to 20° cut at the runway may reduce the drift-control correction needed by half. For the cut is the equivalent of a runway more into the wind.

All of which—that is, the entire chapter—is by way of saying that it appears that hundreds of airplanes damaged each year in the landing roll from loss of directional control touched down with drift. That puts the premium on not allowing a touchdown unless in the last second the nose is pointing in the direction in which the airplane is tracking.

LET DRIFT KNOW WHO'S BOSS

What if we do touch down with drift? After all, there must be hope for lost souls. In a tricycle, if the touchdown with drift is in a tail-low attitude, the nose will tend to yaw toward the runway heading, but in most tricycles this is a small or weak effect. The next thing is to get the nosewheel on the runway, with a slight forward movement of the control wheel, and with the rudder pedals centered. That's when she may start to tip, and if she does, renounce coordination, turn the wheel away from the tipping direction, and give it rudder *toward* the low wing to generate enough centrifugal force to pick the wing up. You may end up with the wing level but the airplane headed 10° to 20° toward the downwind side of the runway, but at least the airplane is now under nosewheel control laterally as well as directionally, and the heading will not be unmanageable on a runway of normal width after a center-line touchdown.

As to the taildragger pilot, rest his airframe, he doesn't tip in a touchdown with drift. He touches down with a running start into a ground loop and quick rudder and brake on the downwind side, opposite aileron, gunning it, and if he does not have a steerable tail wheel, raising the tail a bit for increased rudder effectiveness can give the same recovery rate available to the tricycle pilot, provided he is quicker than the tricycle pilot in getting on the ball.

ENGINE-OUT
TWIN-ENGINE FLYING

Let's look now at the most complicated takeoffs and landings: those involving twin-engine airplanes operating on one engine. Here again the airplane as well as pilot flight characteristics are major factors, but in severely more exaggerated form than in single-engine airplanes. And, of course, the angle-of-attack variables furnished by atmospheric phenomena are still in full force and effect.

The lure of the twin-engine airplane is that it allays one of a pilot's (and passenger's) deepest concerns: a forced landing, either on takeoff or in some out-of-the-way place en route. It appears implicit that if an airplane has two engines, the forced-landing likelihood is completely eliminated and the pilot will continue his flight on one engine and land on an airport without incident. It's a pleasant and reassuring dream and we're getting there but we're not there yet.

The premise starts off well. Most of us had our first engine-out twin-engine demonstration with the airplane at altitude and in cruising flight. The pilot gradually throttles one engine back slowly in order not to cool it too abruptly and cause thermal shock to the cylinders, pistons, and valves. Finally he pulls the mixture control back and then feathers the propeller. He may even ask you to perform the flying while he's doing this.

At cruise speed the rudder is powerful and there's no difficulty in keeping the nose from yawing toward the expiring engine by applying light rudder pressure on the opposite side. When the prop feathers, the airplane imparts a slight feel of acceleration. If the indicated airspeed has been 160 in cruising flight, a little rudder trim and the retention of 75 percent power on the operating engine will end up with an indicated of as much as 130. So, there you are. At 130 it will take a bit longer, but you can go anywhere any single-engine airplane cruising 130 can go.

On this basis virtually all that engine-failure accidents contribute to the engine-out record would be eliminated from the twin-engine record. Happy day!

But it may be a short day if you happen to come across the official pronouncement that over a recent five-year period, twin-engine and single-engine airplanes had almost exactly the same number of reportable accidents from all causes per 100,000 hours flown. You begin to puzzle over what happened to the presumed advantage of having an extra engine.

Next, you may take heart in the NTSB's saying that actually twins have only half as many engine-failure accidents as singles. With two engines twins are bound to have twice as many engine failures as single-engine airplanes equipped with the same engine, so if they have only half as many reportable accidents then the extra engine must make the twin pilot two times times better off with his extra engine.

But then, alas, NTSB pulls the rug from under you: percentagewise the twin's accidents following engine failure are four times more often fatal than those of the single-engine airplane after an engine failure. And NTSB concludes, ominously, that it is four times safer to fly in a single as in a twin insofar as engine-failure accidents are concerned. But that is surely not a conclusion to be engraved in stone. What the four-times figure means to me is that after an engine failure, the single-engine pilot does four times better getting down under control in his inevitable forced landing than the twin pilot does in getting up, around, and down on one engine. Rather than considering a forced landing, as he sometimes should, he concentrates on getting to an airport, or back to the airport.

Single-engine airplanes that are undamaged in forced landings are not reportable accidents, and there are a lot of such successful landings. The equivalent of this is a proportionally even larger percentage of twins that make uneventful landings on airports after an engine failure. The singles, of course, do not get any statistical credit for their off-airport landings, nor do the twins for their also uneventful single-engine landings on airports.

That may be beside the point, but it does emphasize the need for a hard look at how we get our four-times-higher-fatality tag. We get it primarily in engine failures in initial climb and in flubbed single-engine approaches. So we need to examine both the twin pilot's reactions and the performance available to him with an engine failure on takeoff, as well as his approach pilot/airplane performance on one engine. The entire environmental situation is so utterly different, you may wonder why there's any reason for trying to translate back and forth between single and twin statistics.

PUTTING THEORY INTO PRACTICE

The best check-out I ever had in a twin was at Rockwell's Aero Commander Plant at Oklahoma City in November 1971.

The first day and a half were spent covering the airframe, flight controls, and various systems. On the second afternoon Bob Sackett, an old Navy hand and test pilot, a forceful and earnest speaker, told us exactly what would be covered in the flight program next morning.

He wasn't selling Shrikes. The people there either already had bought or else had been employed to fly one. His sole concern was teaching us how to cut the mustard with an engine out, and he said he'd be sorely disappointed if anyone left there without understanding the importance of and really believing in the single-engine climb speeds of the airplane.

We would become familiar with these speeds the next morning by determining what they were ourselves. We'd find that with the left prop feathered and the right engine wide open and the prop in flat pitch, the indicated airspeed would settle down on 130 MPH in level flight. The next step, still going full bore on the right engine, would be to go into a 115 MPH climb, in which the rate of climb would stabilize on 50 FPM. Then we'd get it climbing at 110 MPH and would see 100 FPM climb. Finally the magic number: climbing with 105 indicated the rate of climb would be 300 FPM. That would be the highest single-engine rate of climb anyone could get out of the airplane and the only way it could be gotten would be by maintaining the 105 maximum rate of climb airspeed.

Next, to get a look at the downhill side of single-engine climb performance, we would increase our climb attitude slightly until the airspeed moved from 105 and settled on 100 MPH. This would be the maximum angle of climb speed of the airplane and the rate of climb would stabilize on 275 FPM. Over Niagara Falls in a barrel? Slow just another 5 MPH into a 95 MPH climb attitude, and the airplane wouldn't climb at all but just fly level at that speed.

These numbers were something to remember. Only 5 MPH slower than maximum angle of climb speed and we wouldn't climb at all. Only 5 MPH faster than maximum rate of climb speed of 105 and we'd lose 200 FPM in climb capability. Another 5 MPH faster and the climb would decrease to almost nothing.

Sackett was convinced that where obstructions on takeoff were a factor, it was important right after lift-off to fly at maximum *angle* of climb speed rather than maximum *rate* of climb speed. He drew a tree on the blackboard and flight-path lines to show that with an engine failure right after lift-off with a speed of 100 MPH indicated, the tree would be cleared, but if, after lift-off, the nose was kept down to get 110 or 115 MPH before going into a climb attitude—if an engine quit right there and we went to maximum angle of climb speed, we'd hit the tree halfway up.

I'm sure he's right about altitude being worth more than surplus airspeed in the initial climb stage, and the FAA agrees, but on a reasonably long runway with a clear exit, I feel more comfortable at maximum single-engine rate of climb speed, plus 5. I think most pilots go out of the airport that way, or faster, because in so many twins the attitude climbing on both engines at engine-out maximum angle of climb speed is disturbing to passengers.

Sackett also covered what else would be done in airwork prior to engine cuts on takeoff. Slow flight on both engines with the stall horn just starting blowing would show in 90° 15° bank turns that there was no adverse aileron yaw. Finally, on one engine, we would nibble at V_{mc} to see that even if it took full rudder and that wasn't enough at V_{mc}, banking only a few degrees away from the yaw would keep it straight. V_{mc} would be 75 MPH.

In short, we'd find not only that the airplane would be maneuverable in shallow banks at low airspeed, but that it wouldn't kick up its heels or roll off in a single-engine stall.

Our airwork at altitude would also include some engine

cuts climbing at 105 indicated, with gear down and takeoff flaps. With the cut engine's prop still windmilling and the nose kept straight with rudder, we would be asked to point to the throttle of the cut engine. After that the check pilot would bring up the power to zero thrust setting and the ball would be trimmed to center with the wings level. Sackett mentioned that in low wings the best rate of climb came with the airplane trimmed to fly with the live engine down 5 degrees and the ball slightly on the low side; but this procedure wasn't for the Shrike.

Next morning for my check-out, I drew Leo Broadhead, another of the test pilots, who in mien and thinking was virtually a carbon copy of Sackett. After our airwork Leo explained there'd be a couple of two-engine circuits and thumps, and then the engine-out on takeoff enlightenment would commence. The pattern would be flown at 800 feet and the same pattern speeds would be used, two engines or one. Don't go by that one too fast. Engine out, the airplane would be flown at the same pattern altitudes and airspeeds it is flown with both engines in business. The shape of the pattern would be a little different—it would be a little wider and more elongated rectangle—but the altitudes and airspeeds on each leg would be identical. The only difference would be in the percentage of power available being used to maintain the prescribed airspeeds.

Since the Shrike runs on the ground with the wing at a negative angle of attack, raising the nose slightly on takeoff when it seems about ready to fly is not the answer. A positive rotation into about a 10° nose-up attitude is necessary. But you don't rotate until the airplane is accelerating into the 90- to 95-MPH range. If an engine cuts in the takeoff run with the airspeed below 90 MPH, both throttles must be closed and the takeoff aborted.

With both engines operating, raise the nose just as you go

past 90. When the airspeed hits 105 the airplane will lift off gracefully, and this being the maximum single-engine rate of climb speed, you're soon looking at almost 300 FPM even gear down and with takeoff flaps. In short, with this takeoff procedure you don't become airborne until you have maximum single-engine rate of climb speed, which in the Shrike is the same with one engine as two.

After one takeoff and landing with both engines operating, I was thinking about that 105 first-rung-on-the-ladder figure and anticipating a second two-engine circuit. I was, to put it mildly, startled when an engine cut out in the takeoff run right after I'd seen the airspeed pass 85. At least I didn't stray off the runway in the process of chopping both throttles. Its width helped as well as its 7,198 foot length.

On the next takeoff with 105 indicated and the first indication of a positive rate of climb, an engine was cut. Even though you suspect it may be coming, and no matter what you've practiced aloft, when you find yourself in the lion's cage it is hard to forget how close the beast is and to concentrate on what you're supposed to do to keep him at bay.

Now, let's see. Left foot forward, way forward. That stopped the yaw to the right, but it's also imperative to keep the indicated airspeed on 105, which means winding up with a considerable decrease in the nose-up attitude. In taking care of these two items I let the right wing get down a little and it took more rudder to keep going straight once I got the wing up. I don't want to make this sound too easy, or too difficult, but when it happens unexpectedly in a nontraining flight, it is for sure a time of crisis. For at this point the gear is still down, the ground is near, you've got takeoff flaps, and a windmilling prop, and even with 105 MPH you're starting to lose altitude. Which puts all the chips on feathering the correct propeller, and not being too long about making up your mind which one that is.

Timidly and in a bit of a daze (we weren't even out of the airport yet) I pointed to the right throttle, which was correct, and went back to the all-important attitude/airspeed bit. At this point, and apparently in no great hurry, Leo brought the windmilling prop power setting up to zero thrust, which called for less left-rudder pressure to keep it straight. The airplane started climbing and I started breathing, and then Leo said, "Well?" So I raised the gear and then the takeoff flaps. There wasn't an appreciable increase in rate of climb or trim change and shortly the rate of climb was a steady 300 FPM. Not much, but certainly reassuring about what the airplane was capable of if the pilot could be counted on to function as necessary. We were getting sea-level gross-weight standard-air single-engine performance at approximately 1500 feet density altitude with only two aboard and full fuel. The climb angle, of course, was pretty shallow, but there was nothing higher ahead.

On the next takeoff the right engine was again cut. This time I didn't have to be prodded about the gear and flaps. On the third engine-cut right after lift-off the left engine turned out to be the ailing one. So my good-engine left foot had to become my good-engine right foot. It took somewhat longer to decide that the left engine's throttle was the one to close. In the interim the airspeed got down to 100, which wasn't serious, but another 5 MPH slower and there would have been no rate of climb at all. It's a strict discipline, for that's the path to loss of directional control.

On the last engine-cut in climbout one of the engines started surging. That makes it open season for feathering the wrong prop. I kept walking the rudder to keep it straight and tried to keep extra close check on the attitude/airspeed situation. Finally I pointed to the left throttle. Leo pulled it back a bit and I pushed it back full forward. There was no doubt about that being where the steady power was coming from.

FIRST AND ALWAYS: FLY STRAIGHT AHEAD

Sackett had stressed over and over that after getting the airplane cleaned up and an ailing engine secured, it was imperative to keep it going straight ahead at maximum single-engine rate of climb speed until 400 feet of altitude had been gained. Then, in a shallow bank, the climb could be continued at 105 MPH and when it was time to turn downwind in a 15° bank, the airplane would be at 800 feet above the airport. Which it was each time we turned from crosswind to downwind leg.

Straight out. Don't dare turn. Maybe even better, straight out to pattern altitude, which is recommended in some twin manuals. Having it deeply imbedded in mind that climbing straight out to pattern altitude is imperative should help pilots ignore that powerful impulse to turn back to the airport. In an ill-considered immediate turn-back in a probable 30° to 40° bank in a gusty condition, a gust could, during the crosswind portion of the turn, stall the wing and hand the pilot a combination of loss of directional control, a stall, and entry into a single-engine power spin. Even if the pilot were able to nurse it around in no more than a 15° bank, there would still be a gust problem. An airplane that will climb only 200 FPM full throttle can't maintain altitude on a downwind heading at low altitude in gusty conditions. Every gust caught up with and flown into causes a momentary decrease in airspeed and loss of lift.

The rest of the Shrike engine-out message is relatively simple. The two circuits on two engines and four on one made me wonder why single-engine approaches so often get fouled up. With 800 feet on downwind each time, we'd level off and throttle back the operative engine as necessary to keep the speed down to 110, the same speed used with two engines. It was always surprising how far back the throttle on the one engine would have to come to keep the speed down to 110. There was certainly no power shortage here. Early on Leo reminded me that downwind was a good time to remember that excess airspeed was the undoing of many a single-engine approach.

Halfway or a bit more on downwind we'd put the gear down and, throttle untouched, this would reduce the airspeed to 105 in level flight. Carrying the nose just a bit lower in the 180° turn onto final kept the airspeed on 105. Surprisingly, two more power reductions were required on final. The prescribed final approach speed was 100, and getting down to this, even after takeoff flaps were applied, necessitated moving the throttle farther back on the good engine. With the field unquestionably in reach, full flaps, and for the 95 across the fence still another power reduction was needed. The main difference in flying the approaches at the same altitudes and airspeeds with two engines as with one, is that with gear down and full flaps and both throttles closed, it's easier than easy to get rid of altitude. In contrast, in an approach with one feathered or at zero thrust, even closing the throttle on the good engine doesn't get rid of altitude as quickly as in the two-engine throttles-closed approach because there's the drag of only one windmilling propeller. There's a tendency to overshoot a bit, because of failure to reduce power enough on final on one engine, while following the geometry of a two-engine approach.

Leo's final question taxiing in was, "Now do you feel entirely comfortable with this airplane?"

My answer was yes. And shortly afterward I left for the East for 10 days of delightful and confident cross-country flying in the Shrike. Maybe I wasn't, but I sure felt a lot safer than the pilot must have felt who, in the Commander's earliest days, took one prop off and flew the airplane from OKC to DCA nonstop just to prove that it would fly on one engine.

ENGINE FAILURE—THE FACTS

So much for theory and practice. It's time now to look at theory and fact and see what happens and what will have to change in order that we might rid ourselves of that four-times-higher fatal engine-out rate in our owner-flown light-twin fixed-wing aircraft (under 12,500 pounds gross by FAA definition). And in the process improve our overall rate of reportable accidents to one far better than the present single-engine rate.

I think I know where to start, but first of all let's dispose of that ubiquitous question floating around in pilots' minds: how often do engines quit? Fatal accidents following engine failure or malfunction due to structural failure in the engine or accessories is a rarity, possibly as low as one in each 100,000 flights. But engines also quit for other reasons, and while the unknown figure of engine failures followed by no reportable accident are a credit to our flying skills, we need to think seriously about what those other reasons are. The reportable accident record derives largely from engines either not getting fuel or running out of oil. Obviously a pilot has a degree of control in this area, not only in checking his airplane, but in seeing that his engine and fuel system get proper routine inspection and maintenance.

Now back to business.

TRAINING FOR TWINS—THE NEEDS

One of the first things needing attention in quest of an improved twin-engine accident record is in twin-engine training. Twin-engine pilots fly 50 percent more safely after they get their rating than they do during their training period. In marked contrast, single-engine pilots fly 10 times more safely during their training period than they do later on their own. The single-engine pilot's troubles are more likely to involve errors of judgment about weather and where they go and some of the weird things they try to do with their airplanes. The twin pilot's troubles derive primarily from errors in technique in the takeoff and landing phase, which better training should reduce. And they need more frequent engine-out practice. Twin-engine owner-pilots tend to fly their twins and think about them just as if they were single-engine airplanes. Just get in and go and think only about how seldom engines quit.

In their special study of light twin-engine aircraft accidents NTSB included 15 case histories of engine-failure accidents they considered representative. Following are accounts of the crucial moments in some of the engine-failure cases. The first three cases I shall recount are with an instructor aboard.

In a training flight that was to include a simulated engine failure during takeoff, the pilot requested permission to remain in the pattern for a touch-and-go landing. Upon completion of the touch-and-go the airplane began to climb out and entered a steep left bank and roll, then assumed a nose-down attitude and dived into the ground. Examination showed the left engine to have been at idle speed with propeller unfeathered. The student pilot had 208 hours with three and a half hours in the 310. The instructor had 22,000 hours with 4112 multi and 1500 in the 310.

In a training flight that was to include a single-engine landing with a propeller feathered, the PA-30 Twin Comanche was observed on short final in landing configuration with the right propeller feathered. With the approach speed too high, the aircraft was still airborne halfway along the 2988-foot runway. At this point, power was applied with the left engine. The aircraft started to climb out, gear retracted, and was shortly in a 30° nose-high right turn at an altitude of about 80 feet. The bank continued to the right, the nose rose higher, and the aircraft rolled to a near inverted attitude and dived almost vertically into the ground. The student had 211 hours, 4 of them in the PA-30. The instructor had an estimated 3688 hours multi, and PA-30 time unknown.

Another example. This was the second training flight of the day for the owner of the Cessna 411. After some touch-and-go landings the flight was cleared for a simulated single-engine approach on a different, longer runway. The aircraft was observed in the pattern, gear up, and flying lower than the standard 800-foot pattern. During the turn to final, with the aircraft in a bank of about 45°, the gear was lowered. The aircraft immediately rolled to the right, culminating in a near vertical dive. The Board surmised that the student's steep turn onto final was caused by overshooting somewhat the final approach course and that in the turn, in gusty conditions and at low altitude, one of the pilots may have made a rapid application of asymmetrical thrust with the airplane in a near stalled condition. The owner had 174 hours including 15 in the 411. The instructor had 6,022 hours, 161 multi, and 10 in the 411.

These cases, and there are frequent similar ones, bring to mind a practice common in single-engine flight training which doesn't work in twin-engine training. In a single-engine airplane, past a certain point not so far along in the program, it seems to me that an instructor has to let the student start teaching himself to fly and that the instructor's job is then to let the student go the limit—as long as it's safe—before taking over. In this manner the student gets a vivid picture of the no-nos.

In twin-engine training I think that the margin between the no-nos and what is safe is so narrow that instructors can't afford to let the student go as far in the wrong direction before taking over as they can in a single.

In the foregoing cases the statistical tags were "failure to maintain flying speed" and "stall/spin." With an engine failure on takeoff in a twin the pilot has a different problem from that in a single-engine airplane. Long before he gets down to stalling speed, if he's thinking just of getting up rather than of achieving the only airspeed that will get him up, he trips over the V_{mc} phenomenon. He's slowed down so much that he hasn't enough rudder control to keep the airplane from yawing and rolling toward the ailing engine. Notice, in the foregoing accident accounts, and in some others to follow, how hard the airplanes often try to keep on flying with the nose too high.

It seems we need to put much more stress than we have on how limited is the performance a pilot has to work with in an engine-out situation in a twin. After a failure the nose has to go down promptly to only a few degrees above a level attitude, and from there on, precise maximum rate of climb airspeed control and heading control are all that can keep the airplane flying. At this juncture any attempt to turn back shouldn't even be considered.

Just as a single-engine pilot for the most part has to teach himself to fly, with an instructor along as a life preserver, I think the twin-engine pilot after getting his rating has to recognize the importance of maintaining his proficiency in engine-out situations and be his own life preserver. A biennial flight review may tell him where he's weak, but most likely he already knows that, and certainly the review cures nothing.

KEEPING CURRENT

There are numerous ways in which a twin-engine pilot can improve and maintain his engine-out proficiency. For instance, as we all know, with an engine failure on takeoff, a pilot has to be able to do everything quickly and right. It may sound simple in a classroom, but deciding for sure which engine is ailing is not as easy as it sounds. That's the first place any rustiness can cause havoc.

Whatever a pilot's mnemonic system—dead foot, dead engine; good foot, good engine; airplane wants to turn toward dead engine; set the power controls (throttles, mixtures, prop controls) in the order of your feet—whatever your system, practice it frequently enough so that if, as, and when the unexpected ever happens you are not going to start swimming in glue, to use an old-time CAA phrase.

As another example, in warm weather on downwind, throttle one back, to zero-thrust manifold pressure, or less if solo, and follow precisely the altitudes and airspeeds and flight path of a normal two-engine approach.

Or, occasionally, on a climbout—maybe for politeness' sake you should only do this solo—throttle one back to zero thrust, trim out the yaw, and get accustomed to the airplane's maximum single-engine rate of climb and especially its most probable disquietingly shallow climb-gradient. And keep reminding yourself that what you're seeing is not only the steepest climb gradient available but is obtainable only with the airplane flown within 2 to 3 miles per hour of its maximum rate of climb speed.

WHEN PILOTS FAIL, TOO

Again from the NTSB study: The pilot of a Beech 58 Baron made a normal takeoff, but at 150 feet altitude, due to a mechanical failure, the left engine failed and the aircraft fell to the ground and exploded. The landing gear had not been retracted and the left propeller had not been feathered promptly. The pilot had 2288 hours of which 2012 were multi, 1136 in the 58. The NTSB comment was that, possibly no longer proficient in dealing with power loss on takeoff and caught by surprise, the pilot became indecisive as to what to do first and allowed flying speed to decrease, leading to subsequent loss of control.

A PA-23-160 Apache became airborne after approximately a 1200-foot run, and right after that, with 4000 feet of runway still ahead, the left engine lost power. The aircraft entered a climbing left 180° turn to an altitude of 150 to 200 feet, at which point the nose dropped abruptly, the left bank steepened, and after another 180° turn, the aircraft crashed into the ground. The pilot had 5100 hours flying time, multi and PA-23-160 time unknown.

The 310D departed from a 6600-foot runway, and immediately after lift-off the landing gear retracted and the airplane began a flat right turn, which continued to an altitude of 800 feet. At this juncture the airplane banked steeply to the right and descended into the ground. Loss of the right engine was due to an unwanted feathering, at full power, of its propeller because the engine contained only three quarts of oil. The airplane had not been preflighted. The pilot had 5000 hours, 1500 in the 310D. Total multi time unknown.

Flying at 14,000 feet on a cross-country flight in a PA-30-250 Aztec, the pilot called Blythe, California, radio and reported an oil leak in his left engine. Shortly afterward he reported feathering and announced that he would land at Blythe. After the aircraft entered the pattern, two witnesses observed it too low and too far out for a single-engine approach. About a mile from the field, gear and flaps down, the airplane pitched up and entered a slow climbing left turn. It then rolled left and spiraled into the ground. The pilot had an estimated 400 hours, multi and PA-23-250 time unknown.

Takeoff run of the Cessna 421 was normal and gear was retracted shortly after lift-off. At this time the left engine began to surge. The pilot attempted a 180° turn back to the airport (they don't say which way he turned) but crashed in a field. He said he had panicked and had made no effort to identify the ailing engine or do anything other than try to turn back. He also stated he got below V_{mc} twice and was lucky to get the wings level before impact. The pilot had 3000 hours, 600 multi and 110 in the 421.

The foregoing case histories bring out clearly pilots' responses to an engine failure early in the takeoff phase, in situations in which single-engine performance was available. And you will have noted with gear and flaps down our piston twins can't maintain altitude on one engine, which invites loss of control in an effort to stretch the power glide.

With an engine failure shortly after lift-off we are so desperate for some altitude that we instinctively keep pulling back on the stick, as you have seen, get too slow, and lose directional control.

We need now to consider also the turn-back syndrome with an engine failure farther along in the climbout. Sometimes, once the ailing engine has been shut down, even though the pilot holds the correct climb speed, turbulence or even a slight downdraft can create a sag in climb that makes it look as if the airplane just doesn't have any climb performance. At this point instinct and fear may take over: the airplane is racked into a steep turn back toward the airport. And directional control is lost. Even without the downdraft the impulse to turn back can be triggered not just by the airplane's low rate of climb, but equally by its apparently nonexistent climb gradient.

WHAT ONE ENGINE CAN DO

So much for pilot performance. Now it's time to look at the single-engine performance we have to work with. A single-engine airplane cannot be certificated unless at sea level in standard air at gross and in takeoff configuration it will climb in FPM 10 times its stall speed in MPH. That means at the maximum permissible stall speed for single-engine airplanes of 70 MPH, such a single-engine airplane must climb 700 FPM. In our piston twins our single-engine rates of climb run mostly between 200 and 300 FPM on one engine. And with our normally aspirated engines, single-engine service ceilings seldom exceed 4000 to 6000 feet at gross weight. All in all this is not much to work with.

At Shrike school Sackett had said that before each takeoff in a twin, a pilot should say to himself, "It will probably never happen to me, but if this takeoff should be the one in which I lose an engine, what will be my procedures?" Thus he will review the numbers and routine. But actually we need, away from sea level, to go a step beyond this and make a judgment as to whether following this takeoff we will have any single-engine climb capability at all. If we don't, we'll have to be programed to think in single-engine airplane forced-landing terms and in terms of using the live engine to give us, at maximum single-engine rate of climb airspeed, a low enough sink rate to provide valuable options as to where to put it. Mighty few twin pilots think in terms of single-engine forced or precautionary landing. But I believe those few who do are the ones who get their twins safely on the ground in some incredible places. Flying an Apache one day at Lock Haven, I asked Howard Piper if he thought it would get across town on one engine if we lost one on takeoff. His answer: "You wouldn't even think about that if we were flying a single-engine airplane." Well, no. I couldn't afford to.

Almost any twin-engine pilot can give you the book figure on his airplane for single-engine rate of climb. There's nothing wrong with such figures for comparative purposes. But what counts can be what, say, a 300-FPM sea-level standard-air figure becomes on a hot day at an airport 2000 feet above sea level, at gross.

Let's consider first the single-engine figures of a typical light twin, the Piper Aztec, to see what effect pressure altitude and temperature—in other words density altitude—and weight have in modifying our basic sea-level engine-out rate of climb. We start with, on a standard day at sea level at gross, a figure of 240 FPM on either engine for the Aztec F. Let's say we're at gross, and the temperature is 90°. Setting the altimeter on 29.92, we read a pressure altitude of 2000 feet for our airport. If the temperature were standard at 52° F, the rate of climb on one would be 160 FPM; but at 90°, the figure is 115 FPM. That's not much, and it's in ideal atmospheric conditions—that is to say, with no turbulence.

An aircraft's service ceiling is the altitude at which it can climb no more than 50 FPM. Presumably this is the altitude the aircraft would be able to maintain in turbulence. Since our takeoff today may be in rough air, which is likely with 90° ground temperature, our 115 FPM in smooth air may turn out to be only 65 FPM. The gain in altitude per NM at today's prescribed maximum single-engine rate of climb speed of 88 knots indicated would be only 45 feet. That's so little gain after flying a nautical mile that it would be hard to believe we were climbing. It would take 10 miles of straight-out climb to get out of most of the ground-turbulence effects and to reach even a relatively low pattern altitude.

What would our engine-out performance be from a 3000-foot pressure altitude field on a 90° day at gross? If the temperature were the standard 46° at that elevation, we would have 125 FPM on one, but with today's 90° ground temperature the figure is 75 FPM in smooth air, but only 25 FPM in turbulence. Which would mean that today at gross, we should forget about trying to get up or stay up and think as we would if we were flying a single-engine airplane after an engine failure. Our good engine would be good only for stretching our glide at best single-engine rate of climb speed.

We would, however, have some options today if flying light in the Aztec F. Say at 4800 gross instead of 5200. On that basis, on the 2000-foot pressure-altitude field with standard temperature, we'd have 260 FPM on one, but only 180 FPM at 90°. Or 130 FPM in rough air. That's a lot better than 65 FPM at gross.

At 4800 pounds on a 3000-foot pressure-altitude field at standard temperature, we'd have 160 FPM on one in smooth air, or 110 FPM in turbulence.

From these figures you can see what a help it would be for the pilot of any piston twin to have in mind similar backstop density altitude/weight figures for his airplane when, in those moments before starting his takeoff run, he decides exactly what he is going to do in case of an engine failure on takeoff. In hot weather at gross on a 2000-pressure-altitude field on a 90° day, he might have little more than 65 FPM single-engine performance; and at 3000 feet pressure-altitude and 90°, he'd have, for practical purposes, no single-engine performance. Consequently, at the 3000-foot/90° field, he had better be light by 400 pounds if he wants to keep going after an engine failure on takeoff.

THE NAVY KNOWS

Count on the Navy to know about flying an airplane on the ragged edge. During World War II they bought three all-wood, fixed-gear, Langley 3-to-4-place twin-engine airplanes that had 90-HP Franklin engines with fixed-pitch wood props. They were to be considered as possible twin-engine basic trainers.

After the war they sold one of these to a friend of mine who let me fly it. The airplane had a Navy-prepared manual in it, and for sure it contained The Word about engine failure on takeoff. On one engine, at 2300 pounds and flown at 88 MPH, the airplane would maintain horizontal flight, but under no circumstances should a turn be attempted. In short, keep going more or less straight to the first suitable landing area. At maximum gross of 2450 pounds, an immediate emergency landing should be made, also straight ahead. Presumably these admonitions were for an engine failure shortly after takeoff and at sea level, since to be in the Navy you have to be at sea level. I found that at 2300 pounds gross, with a couple thousand feet and one prop windmilling, it was possible to go as far as 10 miles in search of a suitable field.

So, hear, hear, straight ahead in any twin in which in any given situation it just isn't cutting the mustard. The Navy knows.

KNOW YOUR GLIDE RANGE

What about flying cross-country in the high country, where we may be operating with service ceilings of only 5000 to 6000 feet? After all, following an en-route engine failure at altitude, we're not in the same situation as the single-engine pilot who loses an engine and has to settle for being able to glide only a mile for each thousand feet the airplane is above ground level with a minimum of 800 to 1000 FPM rate of descent at best rate of climb speed.

In the new General Aviation Manufacturers' Association standard owner's-manual format for single-engine airplanes, there's an interesting table with the title "Glide Range." For one high-performance retractable-gear single, you would find that if you were flying at 8000 feet pressure altitude with standard temperature of 30° F and had an engine failure, you could glide under no-wind conditions 5 miles before you got down to a 4000 pressure-altitude ground level, where the temperature would happen to be standard—of 45° F. This would be gear up, flaps up, approximately maximum rate of climb airspeed, gross weight, prop in steep pitch. No wind.

Say, though, that you were in a light twin in the same area and lost an engine at 8000 pressure altitude and had only a 4000 pressure-altitude standard-air single-engine ceiling—how far could you power-glide before getting to the ground? Frankly, after some years in Twin Comanches I should know, but I'd have to go try it to see. I'd guess the answer would certainly be no less than 50 miles.

As an additional line of defense for the beleaguered small twin, I'd like to suggest that our manuals should also have "glide range" charts, for gross and 340 pounds lighter, down to single-engine ceiling.

I once ran across a pilot who'd lost an engine on an Apache flying at 12,000 feet over the high country 100 miles northeast of Phoenix. He got to Phoenix with altitude to spare and landed without incident. He was glad to have had the spare engine. It had gotten him across some mighty rough and sparsely populated country that was well above his single-engine ceiling.

What about turbocharged engines? They are wonderful to have in a twin, but what they add so spectacularly to two-engine performance tends to obscure their effect on engine-out performance in the lower altitudes. Single-engine service ceilings of 12,000 to 15,000 feet in most of the turbocharged piston twins make it sound as if turbocharging must surely eliminate any engine-out performance problems. But that's not the way it is.

Where the same model airplane is offered with normally aspirated or turbocharged engines, as in the case of the Aztec F, the performance tables show that the turbo model has a little less engine-out rate of climb at sea level at gross weight. This is because our turbocharged general-aviation twins do not have an intercooler between the compressor and the intake manifold, and consequently the hot, thin mixture going into the cylinders is less powerful than a cooler, denser mixture, a factor which is not quite compensated for by the maximum manifold pressure allowed.

For instance, the Aztec F's nonturbo 240 FPM figure at sea level becomes 225 FPM in the turbocharged model. And on a 90° day at sea level, 210 FPM. Which is not a serious loss. But, of course, it puts a bit more pressure on a pilot in a low-altitude failure on takeoff because of the still shallower climb gradient. Where the Turbo Aztec F pilot starts getting his money back is at the point where the standard model starts playing out on one engine. On a 3000-foot pressure-altitude 90° day at gross, he has a 170 FPM rate of climb on one instead of 75 FPM. With an engine failure on climbout from Denver on a 90° day, he'd have approximately a 150 FPM rate of climb at gross climbing with 83 KIAS. Without the turbos, on this day in Denver the best he could manage on one, if he kept his

86 KIAS, would be a stretch of the glide in a possibly most useful, 30 FPM rate of descent.

To state the case in general terms, the Turbo Aztec F pilot does not have to deal with the standard-model pilot's 3000-foot pressure-altitude high-temperature single-engine numbers until he gets to approximately a density altitude of 10,000 feet. And at gross at standard air temperatures he has a 17,000-foot single-engine service ceiling. So I think turbochargers are worth having even aside from the way they keep takeoff runs rather uniform throughout the year and give a hefty two-engine rate of climb, 1350 FPM for instance, in the Turbo Aztec F at 12,000 feet density-altitude at standard temperature.

How can the piston twin engine-failure accident record be improved? Better engine-out performance and better pilot training would head most anyone's list, and it did the NTSB's.

The NTSB had an interesting table of total accident and engine-failure accident rates according to power loadings. The airplanes with the higher power loadings had more total accidents and also engine-failure accidents than those with lower power loadings. Because, NTSB reasoned, the higher the power loading the less power available for climb on one engine. The life preserver they proposed is lower power loadings for increased single-engine rates of climb. Or in other words, bigger engines.

While it is logical enough that the pilot needs more rate of climb on one, I do not think having it would improve the engine-failure-on-takeoff accident probabilities unless along with the increased single-engine rate of climb came a comparable improvement in climb gradient. As it so often turns out, with bigger engines the maximum rate of climb speed at the increased weight is so much higher than before that even though you may climb faster you also get there sooner, and a mile out after an engine failure on takeoff, you're no higher than before. It is climb gradient, not just rate of climb on one engine, that needs to be improved.

This lack of effective improvement in climb gradient despite an increase in power occurs because of an economic factor the statistics do not sense. When the power goes up, consumers and their accountants want something more than increased single-engine rate of climb. Maybe an extra seat, more fuel, and an increase in cruising speed. Otherwise they'd be buying an airplane with a higher seat-mile and ton-mile cost of operation. So the gross and stall speed go up as well as maximum single-engine rate of climb speed, and there's little or no improvement in single-engine climb gradient.

Trying to sell a higher-powered version of a given model airplane that would have maybe a bit more speed but considerably less useful load and range in quest of better single-engine climb gradient would be like trying to sell a boat owner $100 life jackets when he assumes or may have found that the $10 ones he has work well enough. Safetywise it is hard to sell something that will probably not ever be used when the cost is as high as it is for bigger engines.

Which brings us to an important and as yet unanswered question of how much is enough? How much single-engine rate of climb and climb gradient do we need and would be willing to pay for?

Let's have a look at where we are now in our free-market environment. At gross at sea level in standard air, the Cessna 310, with 285-HP engines, climbs 1662 FPM on two, 370 FPM on one at 107 knots, and has a climb gradient of 207 feet per NM. The Cessna 402C, with 325-HP engines, climbs 1450 FPM on two, 301 FPM on one at 104 knots, and has a climb gradient of 175 feet per NM. The 421 Golden Eagle, with 375-HP engines, climbs 1940 FPM on two, 350 FPM on one at 111 knots, and has a climb gradient of 189 feet per NM.

Across the way in Wichita at Beech, the 58 Baron with 285-HP engines, climbs 1660 FPM on two, 390 FPM on one at 100 knots, for a climb gradient of 234 feet per NM. The 58 TC Baron, with an 800-pound higher gross and 325 HP engines, climbs 1475 FPM on two and 270 FPM on one at 115 knots, for a climb gradient of 151 feet per NM.

I hope you will join me on the back row in the balcony and stand in a cheer for the new little Duchess: with its 180-HP engines, it climbs 1248 FPM on two, 235 FPM on one at 85 knots for a climb gradient of 166 feet per NM. Which, incidentally, is 2 feet more per NM than the piston-engine flagship Duke delivers with its 380-HP engines, 1601-FPM climb on two, and 307 FPM on one at 112 knots. I think the Duchess shows there's hope for the lower-powered small twin.

Across hill and dale at Piper, the Seneca II, with 200-HP engines, climbs 1340 FPM on two, 217 FPM on one at 88 knots, for a climb gradient of 148 feet per NM. South, midst square miles of alligators, at the Florida Piper plant, the Navajo, with 310-HP engines, climbs 1445 FPM on two, 245 FPM on one at 94 knots, for a climb gradient of 156 feet per NM. Its stretched version, the Chieftain, with 350-HP engines, climbs 1390 FPM on two, 230 FPM on one at 106 knots, for a climb gradient of 130 feet per NM.

Let's stand and cheer again, for the new little Seminole with its 180-HP engines, which climbs 1340 FPM on two, 217 FPM on one at 88 knots, for a climb gradient of 148 feet per NM.

All the foregoing figures, as mentioned earlier, are for gross weight at sea level in standard air. At higher elevations and temperatures, they would decline in just about the same proportion as the figures for the Aztecs, which were given earlier.

How does the other half fly?

Among the turboprops, the King Air C90, with 550-shaft-horsepower engines, climbs 2520 FPM on two, 818 FPM on one at 107 knots, for a single-engine climb gradient of 178 feet per NM. The Super King Air, with 850-shaft-horsepower engines, climbs 2868 FPM on two, 938 FPM on one at 122 knots, for a climb gradient of 363 feet per NM. Cessna's Conquest, with 635-shaft-horsepower engines, climbs 2435 FPM on two, 715 FPM on one at 120 knots, for a climb gradient of 357 feet per NM. Rockwell's Turbo Commander 690B, with 700-shaft-horsepower engines, climbs 2821 FPM on two, 879 FPM on one at 115 knots, for a climb gradient of 462 feet per NM.

Among the jets (hold your hat), the Citation II, with 2500-pound-thrust engines, climbs 3250 FPM on two, 910 FPM on one at 152 knots, for a climb gradient of 423 feet per NM. The Gates Learjet 24F, with 2950-pound-thrust engines, climbs 7100 FPM on two, 2050 FPM on one at 220 knots, for a climb gradient of of 613 feet per NM. A 613-foot climb gradient sounds mighty good, but at what a price!

Again, the question: how much, or how little, is enough?

From the regulatory standpoint, if a light twin weighs 6000 pounds or less and has a stall speed of 70 MPH or less, gear and flaps down, it is not required to have gear and flaps up, any single-engine rate of climb at all. Light twins that gross over 6000 pounds or with a stall speed in excess of 70 MPH, at 5000 feet must, with one prop feathered and gear and flaps up, climb 2.7 percent of the square of their stall speeds. This would mean that an airplane stalling at 70 MPH would have to climb 114 FPM at 5000 feet. This requirement would also apply to the under-6000-pound or under 70-MPH stall-speed twins that are used for air taxi or commercial operations or are capable of hauling 10 or more passengers.

The turbocharged Seneca II and the turbocharged version of the Seminole would easily meet the 5000-feet air-taxi climb requirement. The unaspirated engine Seminole would not, since its single-engine ceiling is 4100 feet. The Duchess, also normally aspirated, even with its 6100-foot single-engine ceiling probably would not meet the requirement.

I once asked Franklin T. Kurt of Grumman, one of general aviation's most ardent supporters over the years, what it would take to get good single-engine performance in a small twin. His answer was that the airplane would have to be so overpowered that it would be uneconomical. I suppose that's still true, and certainly all the foregoing piston single-engine rate-of-climb and climb-gradient figures suggest that more power would surely help.

But the way the Duchess, Seminole, and Seneca II hang in there with climb gradients that survive the pressures of the marketplace tempts me to believe they are here to stay. NTSB says that the two principal factors affecting safe single-engine flight of light twins are airspeed and power. I think there are not two but three principal factors, namely airspeed, power, and wing loading. As power goes up in the twins, stall speeds are allowed to go up to get better cruise speed. The single-engine maximum rate of climb speed also goes up, which flattens the climb gradient. For better single-engine climb gradients we need not so much higher single-engine rates of climb as we do a steeper angle of climb. It seems to me that the route to this improvement is not just more power but also lighter wing loading.

From the marketing standpoint that's a desperate hypothesis. If the increased wing area on a given airplane decreased its cruising speed 10 to 15 knots, you know good and well whether a slow but better single-engine climb-gradient model or a faster sport model would dominate the market. Alas. And shades of the Buhl Pup of yesteryear. It was offered with a long wing for those who wanted to land slowly and a short wing for those who wanted to go fast. The latter was what sold.

The seat-mile costs of our airplanes have to be competitive

with those of other means of transportation, and the slow boat to China doesn't accomplish that. Bob Buck once told me that a high-up engineer at one of the major airline equipment factories told him that even if the single-engine 70 MPH stall-speed limit on single-engine airplanes had been applied to twin-engine aircraft, we'd still be flying 600 MPH and off 5000-foot runways. The way to accomplish this simply would have been found. Such an airplane would probably not be cost effective, though, compared to other means of transportation. I think we're in the same stall-speed boat with our piston twins, in which I'd still rather have an engine failure over, say, Manhattan, and a lot of other places, than I would in a single-engine airplane. I think the twins' overall benefits more than offset the engine-failure-on-takeoff dilemma. Again, Buck: "The secret of safe twin-engine flying is good engines."

DOUBLING-UP FOR RELIABILITY

Along that line, and though he was thinking more of single- than twin-engine airplanes, Fred Weick once proposed, after an extensive study of the causes of engine failures, an engine with dual fuel systems that would operate simultaneously at 50 percent capacity and either one of which could supply fuel for full power if the other failed. This, he felt, would eliminate a lot of engine failures. He also proposed that such an engine have dual oil passages throughout the engine and dual oil pumps. These features would add to the cost and weight of engines but would eliminate by Fred's calculations possibly 90 percent of all engine failures. No sale. It would cost millions to develop a family of engines of this type, and whether they would sell at the necessary higher price would be highly uncertain. Thousands upon thousands of pilots have never even had an engine miss. How would you convince them they need a better, more expensive engine?

Certainly I do not need to pass the hat for our aircraft manufacturers, for they are far from destitute. They are a shining example of the virility of free enterprise. They couldn't survive building something people wouldn't buy, for whatever reason. But I do think they often make more serious efforts in the safety area than they're given credit for.

Remember the Jato bottles? The old Twin Beech was no record breaker on engine-out performance, but with its good engines there were few complaints. Even so, at Beech they offered a neatly installed Jato bottle atop each wing in the back of each engine nacelle. With an engine out on takeoff the corresponding Jato bottle could be fired and the airplane would walk right on up 1000 feet or more at close to its normal two-engine rate of climb. Quite a few installations turned up on corporate 18's, but initial cost, extra weight, the need for periodic rejuvenation of the bottles, and the fact they were so rarely used, resulted in their eventual disappearance.

SAFETY IN CENTER-LINE THRUST

Then at Cessna, the 337 push-pull. It deserves a place in history on a par with the Ercoupe effort to make a safer airplane. First of all it eliminated what many regard as the first and highest hurdle in the engine-out situation by means of its center-line thrust. Lose an engine and it would keep going straight and level—more slowly, of course, or with a much lower rate of climb, but straight—in great contrast with the conventional situation where the airplane wants to roll in a skidded turn with asymmetrical thrust. And there was no doubt about its ability to climb on one.

On my first Skymaster flight with Bill Thompson, an engineering test pilot at Cessna, we taxied over to the long runway at McConnell Field, he feathered the front prop, and I took off. It climbed a good 250 FPM even with its fixed gear. Later this got up to 300 FPM with a retractable version, and in the turbo model to 335 FPM on the front engine and around 350 on the rear. They even had a pressurized version with 225-HP instead of 210-HP engines. It climbed 1170 FPM on two, 335 on the front or 375 FPM on the rear engine. The climb gradient on all these models ran from 205 on the first to 253 feet per NM on the pressurized version.

Sad to say, the 337's engine-failure accident record was a great disappointment. Some of the trouble came from the failure of pilots to bring up the power on the rear engine first at the start of a takeoff to be sure it was running. But mainly I think the 337's record reveals that even without the bewildering distractions of the yawing and rolling caused by asymmetric thrust, an engine failure in any twin does more to stop the wheels turning in a pilot's head than such failure would in a single-engine airplane. In the latter situation it's at least quiet.

The Ercoupe did not survive because pilots do not like to be overprotected. There might have been a slight element of this in the 337's rejection. Some feel that its unconventional appearance was its principal sales handicap. My guess is that what undid it marketwise was the fact that in a comparable conventional twin you could go just as fast as the 337 on 100 horsepower less—which meant more economically. What price safety!

Until someone develops a more feasible combination of power loading and wing loading for the small piston-engine twins and figures out how to sell it, we are going to have to live with our shallow engine-out climb gradients.

That suggests better and more intensive pilot training. Officially, that nearly always leads to bigger doses of what isn't already working. Such as V_{mc} demonstrations at quite low altitude in normally aspirated twins so as to get maximum power on the operative engine and thus maximum yaw effect. All one has to do in this situation is hit a good asymmetrical bump to get a roll-off and dive into the ground. Or often, especially on Air Transport Pilot flight tests, a call is made for a balked landing with the airplane flared and gear and flaps down. Just as the pilot gets both throttles forward, an engine is cut by the examiner with a resulting cartwheel and destroyed airplane. Once full flaps are applied on final in a piston twin, there's no point in attempting a single-engine go-around unless there's enough altitude to trade for acquiring and maintaining maximum single-engine rate of climb speed while getting the gear and flaps up. Unless such a trade-off is possible, the only alternative is to go ahead and land willy-nilly.

PRECISION AIRSPEED CONTROL

"Failure to maintain flying speed" is practically a rubber stamp on engine-failure accident files. This is not a statistical label indicating an obtuse pilot; its real meaning is that it isn't easy to maintain flying speed when a pilot has so much else to do after an engine failure.

To maintain flying speed we have to be able to exercise precise control of our airspeed indicator, which is a rate instrument. There are two ways to control it precisely. One is by a cut-and-try attitude-adjustment method. By looking over the nose or at the artificial horizon we pick an attitude and hold it and then check to see what's happening to the indicated airspeed. If it's steady but too low, we lower the nose a bit, give the airplane time to accelerate, and then check the indicated airspeed again. And so on, until we get it where we want it as related to a given stick or elevator position.

The other method of control of indicated airspeed, which is more useful in correcting large excursions from desired indicated airspeed, is to fly the trend and rate of movement of the airspeed needle. If the needle is on 60 knots but we need 75 for, say, maximum single-engine rate of climb, we move the stick forward a bit and then a bit more if necessary to get the needle moving toward 75. If it starts moving rapidly, we have to move the stick back enough to decrease the rate of movement of the needle. With the needle moving at a slow rate toward 75, we can stop it on or close to 75 with a final slight movement of the stick forward. And vice versa on stick movements and positions for getting rid of too high an indicated airspeed.

Note that stick or elevator positions have been mentioned, not stick pressures. If you have the time to spare to figure out what the changes in stick *pressures* would have to be to accomplish the foregoing airspeed corrections and control, you'll have to start with the airspeed for which the trim tab is set. Happy labyrinth! And don't get to chasing the airspeed needle with stick-pressure changes in the wrong direction!

There is a solution to this critical problem, which requires neither skill nor practice. It is the left-right, slow-fast needle of an angle-of-attack indicator manufactured by Safe Flight Instrument Corporation of White Plains, New York. It is mounted atop the instrument panel just to the right of the pilot's line of sight over the nose. When the needle is straight up in a climbout, the wing is at the angle of attack that produces the maximum lift/drag ratio, or, in short, the maximum rate of climb speed. The needle is free of lag, and moving it with the elevator control is as positive as if there were a direct mechanical linkage to the needle. So much for the money so simply!

The airspeed reading on each climbout or approach with the needle straight up pointing to a maximum lift/drag ratio mark won't be the same, but the centered needle always gives maximum performance because it automatically considers variables such as the weight of the airplane, density altitude, power available, and, of great importance, the effect of turbulence. The angle-of-attack indicator makes all these necessary computations by sensing the movement of the stagnation point on the leading edge of the wing—the point where the air separates to go over and under the wing—in order to get the maximum performance available. You can see what flying this easily controlled and centered needle is worth when at the high-density-altitudes the single-engine rate of climb is only a fraction of the sea-level book figure and flying only a few miles too fast or too slow can eliminate it entirely. No pilot could make a better investment for maintaining not just flying speed, but the exact and variable flying speed needed in any given weight, density altitude, power, and turbulence situation. The owner's manual contains the necessary charts for all these variables except turbulence, and that can make a considerable difference in determining a correct higher single-engine maximum rate of climb speed. With the angle-of-attack indicator you don't have to ask. It tells you. And, along with the angle-of-attack indicator, if times would permit, I'd like to have an instantaneous vertical speed indicator for an immediate report on how well I was doing.

Finally, something that isn't talked about in engine-out twin-engine flying. All of us, flying singles or twins and normally climbing around 1000 FPM, have seen days on which the airplane would stagger along after takeoff at only half its normal rate of climb. This could be due to turbulence, or to taking off in an area of subsidence, or from an overrunning tailwind, or from climbing out in an oversize downdraft. But whatever the cause, the minus 500 feet per minute or so in rate of climb means that our twins with their 200 to 400 FPM climb capability on one engine are going to settle on these rare occasions rather than climb at maximum single-engine rate of climb speed.

This means that no matter what the pilot's pretakeoff estimate of single-engine performance should be if things are done right, or regardless of what the book says about the airplane's climb capability under given circumstances, it's going to be up to the pilot to appraise the actual situation after any engine failure on takeoff. It's either climbing or it isn't, and if it isn't then we have to program ourselves to concede that the twin is not invulnerable to a forced or precautionary landing. Even though the situations are rare, many of our troubles come from trying to stay up rather than using the power we have left to pick the most favorable place to get it on the ground, and decelerating.

The low-powered small twins? They are on the low end of the single-engine performance scales, but so long as airplanes like the Duchess and Seminole can sail along indicating 120 MPH with one prop feathered and the operating engine at only 75 percent power, they are going to sell. In these airplanes, as in any other twin, there's peace of mind cruising over forbidding terrain, or a metropolitan sprawl, or flying at night, or over water, or on instruments, because in the enroute phase one can think comfortably of engine failure in terms of going to the nearest suitable airport. While it's true that forced landings are not statistically significant as an accident cause, they are a great psychological burden of which the twin-engine pilot is free 99 percent of the time. The 1 percent takeoff time exposure we can handle a lot better than we do with some basic rules and a reasonable amount of zero-thrust single-engine practice.

As it is, with an engine failure on takeoff, even with center-line thrust, our first problem is a psychological one: all we can think about is getting back to that airport behind us. We forget that an airplane can't turn without losing airspeed unless additional power is available, and in this situation it isn't. So it is inevitable that we lose control in our attempted panicky turn-backs.

There was a time when the phrase "professional pilot" described a person flying for a living. Today being a professional pilot more often implies advanced training, as for an ATP, and there are outstanding advanced schools such as Flight Safety at Vero Beach, and some factory schools for pilots of the company's higher-performance aircraft. I think the private owner-pilot of a twin needs to become, even if only self-educated, at least a semiprofessional pilot.

FOR "OLD-PRO" ENGINE-OUT PROFICIENCY

He should have enough engine-out zero-thrust practice, regularly, to assure that identifying a failing engine and getting the prop feathered, the rudder trimmed, the airplane cleaned up, the ailing engine secured, is an easy reflex action. Along with instantly going to and hanging on to the maximum single-engine rate of climb speed and small nose-up attitude.

From there on I think the saving grace is a realization that in gusty conditions an airplane that is climbing poorly at low altitude, which almost any of the smaller twins is going to be, is going to climb best if kept headed into the wind until pattern or higher altitude is reached. Flying into the wind in gusty conditions, the airplane has the benefit of kinetic energy absorbed from the gusts, which improves climb performance. Such an airplane on a downwind heading at low altitude may simply slowly settle to the ground at maximum rate of climb airspeed.

So, after the first routine automatic household chores of getting into a proper climb on one engine, the second thing is to keep going straight for a few miles. Then level off, get the speed up a bit, and make a shallow turn downwind. And from there on fly the same speeds, pattern, and configurations used in normal two-engine approaches.

Normal two-engine approaches? We'd be well ahead of any evil day if we acquired the habit of keeping these approaches steeper than most twin-engine pilots do with low enough power on both engines to keep the approach steep enough so that with an engine failure at any point on final we'd without question be able to make it to the airport on one engine. This doesn't invite overshoots, because with gear and flaps down and two windmilling props we are able to correct an approach path that is too high. As suggested earlier, this type of approach practiced with, occasionally, zero thrust on one engine would make it a routine rather than emergency situation.

Along with the thorn there's a rose: in corporate twin-engine flying, much of it still in piston twins, the fatal-accident rate is right at one million hours flown per fatal accident. That is roughly eight times better than we do with our owner-flown small twins. Some of the difference is due to better single-engine performance, but I believe most of it reflects the skill and self-control of the pilots, who are content to let the airplane do its thing on one engine. Which means, simply, keeping the angle between the relative wind and the mean chord line of the wing right at 7° in stabilized flight in climb, approach, and in turns, or in other words sticking within a mile or two of maximum rate of climb airspeed in these situations. That's what flying and flying safely is all about, and most especially in a twin on one engine.

LOST CAUSES

I have a feeling for lost causes in quest of safety. As mentioned earlier, the list is a long one, but surely it would be unfortunate if that stopped all further effort to solve some of the more critical pilot/airplane problems, even though few such efforts have been successful. The failures to achieve marketing acceptance occurred not because the devices failed to do what they were designed to do, but because of some often obscure negative reaction of the prospective consumers, to which, of course, they were fully entitled.

So, you might say, in memoriam:

Wing tip leading-edge slots, which cost a few miles cruise but significantly improve lateral stability at the stall. Stall-resistant, spin-proof airplanes, like the Ercoupe and Skyfarer, the former also automating the crosswind landing. Tactair's automatic speed/trim control, which would make an airplane fly, power on or power off, within a few miles of whatever speed it was trimmed to fly.

Mooney's full-time wing leveler, which necessitated holding a wing down in a turn, as we do with pressure on the steering wheel to keep an automobile in a turn. This device, in effect, provided positive spiral stability and, unless overridden by the pilot, prevented the spiral dive. And it saved many a Mooney owner. But apparently pilots do not like positive spiral stability and prefer airplanes that will end up in a spiral dive if turned loose.

In his post–World War II Culver V, Al Mooney had his Simplifly lever. It could be set for takeoff, cruise, or approach and operated the elevator trim tab. Its purpose was to keep the airplane flying at a safe airspeed in these phases of flight without any effort on the part of the pilot. For instance, if it got a bit slow in a climb, the nose would snap down to level attitude before the pilot could get to it. It provided pronounced longitudinal stability, which, of course, could be overridden if the pilot felt that necessary.

More recently Fred Weick, in a research project in conjunction with the University of Michigan, made a more sophisticated effort with a sort of floating down-and-up spring in the elevator control system to increase longitudinal stability by taking the trailing float out of the elevator with trim setting. He also took care of CG movements by being able to set his trim lever for various combinations of passengers front and back in his Cherokee, and also baggage-compartment load. The device worked all right, but in trying it out on pilots, the University found their reaction negative. Fred is now noodling a mechanical lash-up which would cause an airplane to fly at a constant angle of attack instead of at a constant airspeed, but is not so sanguine about its acceptance.

I was told once at Teledyne's at Charlottesville that the Navy was having serious trouble with too hard landings in the heaviest aircraft they were landing on carriers, and even, sometimes, with damage to the deck. Installation of Teledyne's quite sophisticated angle-of-attack indicators eliminated the problem completely. So far as general aviation is concerned, let's add Doc Greene's Safe Flight angle-of-attack indicator to the list. Especially for the twin-engine pilot on one engine. Thirty years ago I flew with him on a gusty, hot day in his E-18 Twin Beech. With the right engine throttled back to the prescribed zero-thrust setting just as we went out of Linden Airport in New Jersey, he had me fly for three minutes at maximum single-engine rate of climb speed. Then the flight was repeated, this time flying with the angle-of-attack indicator needle kept on the maximum lift-drag ratio mark. At the end of three minutes on this second flight, we had, instead of the 450 feet gain in altitude of the first flight, nearly 600 feet gain in altitude.

I flew a Twin Comanche a year with a left-thumb button on the control wheel, which was a joy. Anytime it was necessary to hold back or forward pressure on the control wheel to maintain an attitude, holding the button down would get the airplane trimmed and the trim motor would shut off automatically thanks to its spring-loaded toggle switch. Pilots did not like the lash-up because it would trim off all except about a pound of stick force necessary to hold the desired attitude, which necessitated hand trimming if that was not acceptable. It certainly lightened the work load at otherwise busy times. My only reservation was what might happen if the button were held down with the airplane in a spiral dive, but I didn't try that.

EVERY LITTLE BIT HELPS

The Jato bottle. The crosswind gear. Propeller unfeathering accumulators, which are still available but seldom bought. I've never felt comfortable feathering a propeller without one foot on a good-size airport. This device can give an instantly windmilling propeller in a restart. In my own, and in a number of demonstration flights in other aircraft, I've experienced some drawn-out seemingly chancy restarts of an engine with its propeller feathered. On occasion I have gotten a guarantee from my home tower of number one to land and landed with a propeller feathered. Restarting on the ground with a propeller feathered gives an exceedingly rough start-up after a lot of cranking. I understand that at some of the advanced twin-engine training schools they use unfeathering accumulators and consequently can safely give a lot of approaches with one engine feathered. It adds a bit to the float after a normal single-engine approach and flare. I'd like to see more unfeathering accumulators on demonstraters.

And for good measure and with some timidity I'd like to add a limbo candidate of my own now in process: an airspeed indicator with a built-in accelerometer, which will operate a second hand on the standard airspeed indicator face through a concentric shaft.

This second hand, painted red and labeled G-STALL in stabilized flight conditions, that is, in a normal climb or glide, will point to the advertised stall speed of the airplane. In a turn, or pull-up, or pull-out from a dive, the G-STALL needle will move toward the indicated-airspeed needle and point to the stall speed, with proper calibration for each aircraft, under the G-load conditions existing in the turn, pull-up, or pull-out. In those situations the pilot will be able to tell from the distance between the two needles how much margin he has before the angels start singing, and will be taught how easily he can increase the separation of the needles under G-load conditions simply by a slight forward movement of the elevator control.

LET'S FLY

Let's fly a few times around the patch with this background thought: Why should anything untoward ever happen when we're taking off and landing? In our cross-country or en-route flying most of the troubles are ascribed primarily to gross errors in judgment. In our takeoffs and landings the premium is more often on our basic flying ability. Let's look closely at what we're supposed to be able to do in our airplanes.

Our most common motivation as general-aviation pilots is a desire to be free to go in whatever direction we wish, fast, economically, comfortably, and safely. In order to do that we need to provide three things: angle-of-attack control, heading control, and altitude control. That's what gets us where we want to go. It also gets us up and down.

In the exercise of these control functions during takeoffs and landings there are commonly many distractions, and for one reason or another our airplanes are often not at all cooperative.

KNOW YOUR MACHINE'S CHARACTERISTICS

Just how a particular airplane *wants* to fly in any given situation is not under the control of the pilot. But how that pilot *makes* it fly is. Given its head an airplane wants to fly a certain way because of its design characteristics, its stability and control characteristics, and the setting of its trim tabs. The pilot can't change any of this except trim. When he's busy, all he can do is supply the control forces necessary to stabilize the airplane in the attitude he wants it to maintain. Which can at times call for a heavy hand and strong foot.

The better we know our airplanes' flight characteristics and trim responses, the better we can know what's coming next. The main thing, though, is that we should not let these at times obstinate flight characteristics and trim quirks alarm or confuse us. And that we should by no means let them even momentarily take over our function of administering the trinity of airspeed, heading, and altitude control.

Cleared for takeoff? If it's a taildragger, we take a good look over the nose once on the active and file a mental picture: this is the attitude we want to duplicate just before touchdown in landing. If it's a tricycle, we also take a look over the nose: the picture we want on touchdown will require a more nose-up attitude. In either gear configuration this look can also give a clue to where we'll to have to look for best judgment of altitude after round-out. All too often the best place is not over the nose but around and ahead on one side of the nose. It's also helpful to use the same seat height and position adjustment on each landing. Some airlines assure this with a pawnbroker's sign in reverse mounted at the top of the windshield. Pilot and copilot adjust their seats until the ball on their side is lined up with the forwardmost of the three balls.

Away we go! Full bore. If the so-called torque effect at the start of the takeoff run bothers you, bringing up the power a bit on the slow side will provide an opportunity to get a better feel of how much right rudder pressure it's going to take to keep it straight. And, of course, this pressure diminishes as the takeoff run progresses.

In a crosswind takeoff we may want to run to a higher-than-normal speed before rotating to be sure of not touching back with drift, but the nosewheel on some of the tricycles gets sensitive enough to invite overcontrolling if the airplane is kept on the ground too long.

In a tricycle with a free-swiveling nosewheel a takeoff in a strong crosswind from the left may require not only full right rudder well into the run but some right brake in addition until the true airspeed is high enough to create sufficient air load on the rudder to make it an effective heading control.

CLEARED AND ROLLING

Over- or undercontrolling in the takeoff run can be embarrassing, but it is seldom of any great consequence. If we're flying one of the light twins with counterrotating propellers, we're home free from any torque problems in the takeoff run and even in the climbout.

In a taildragger, tail up to level attitude early for best acceleration. At the prescribed takeoff speed the airplane will lift off much more nicely than a tricycle does.

If we're flying a tricycle, there may be a note in the owner's manual about lightening the load on the nosewheel slightly at some particular speed. Nosewheels can take quite a beating on even a slightly rough runway at high speeds. But after rotating, the tricycle doesn't lift off on its own the way a taildragger does. In the tricycle the mains are behind the CG, dragging, and the up-elevator it takes to hold the lift-off attitude is all of a sudden too much when the airplane becomes fully airborne and is relieved of the drag of those rearward wheels. So we may pitch up a bit—especially in a forward CG takeoff in many of the six-placers—because it took extra elevator up-travel in the takeoff run just to hold the nose up off the runway. The pitch-up calls for at least a small reduction of the back pressure on the control wheel, but, of course, not enough to permit a touchback.

GEAR-UP IN INITIAL CLIMB

Some of the retractable-gear tricycles seem to lose their eagerness to climb just after lift-off. If so, just hold the nose steady for a moment or so until it gets its breath. Premature or excessive rotation for lift-off can sometimes give a vigorous response in climb—until the airplane gets to the top of the ground cushion. Then the climb enthusiasm diminishes and the nose needs to be lowered a bit from the lift-off attitude. In a normal takeoff in a tricycle it is helpful to remember the nose-up attitude when it lifts off. That's the least in the way of nose-up attitude you'll want to see in a tail-low touchdown.

The instant we're airborne, we move from the relatively uncomplicated world of the road runner to that of the pilot, and we have to really start flying. Simultaneously precise airspeed and heading control must be exercised. The latter has become more complicated than it was in the ground run because now a wing being down can affect where the airplane wants to go. Whether heading control or angle-of-attack control should have priority might be debatable. But let's consider angle of attack first because in any stabilized climb or approach, the angle of attack has to be controlled precisely. We can't see angle of attack for lack of a piece of string out front such as the Wright Brothers used, so we have to use a somewhat devious indication, which is airspeed.

Some pilots prefer to rotate to a conservative lift-off attitude and wait for the airplane to accelerate in the climb to its best rate of climb speed. Others prefer to get a positive lift-off and then hold down a bit for a shallow initial climb and let the airspeed build to maximum rate of climb speed before really starting up. They don't feel comfortable close to the ground climbing with an airspeed lower than they'd be happy with on final approach.

One way or the other, once you get to maximum rate of climb speed after takeoff, you want to keep the indicated airspeed on that figure, for two reasons: first, the sooner you get higher, the more options are available with any sort of engine malfunction; and second, maximum rate of climb speed not only gets you up the fastest, but at that speed you are comfortably out of reach of what a gust or turbulence can do to increase angle of attack without any change in attitude of the aircraft. Gusts and updrafts don't do anything serious to the pilot climbing at maximum rate of climb speed. But they can stall a wing that is climbing well below maximum rate of climb speed. The nose can drop precipitously in a too steep climb in severe turbulence, and this can cause a pilot to attempt to jerk it back up to the selected climb attitude, when, instead, a bit of altitude should be traded for more airspeed.

THE "CUT AND TRY" TECHNIQUE

How do we maintain maximum rate of climb speed precisely? By cut-and-try attitude adjustments. From experience we learn that the nose rides at a certain height in a full-power climb at maximum rate of climb speed, or a little lower in turbulence. Or we may start with the dot in the center of the artificial horizon raised to the 10°-up bar. Then we bide our time a bit and check the indicated airspeed. If it's lower than we want, the nose is lowered slightly to a new cut-and-try attitude. Or if the climb is at too high an airspeed, we increase the nose-up attitude a bit and hold it there to see where the indicated airspeed will settle down. Pushing or pulling on the control wheel until the airspeed needle is on the desired maximum rate of climb number leads to the porpoising climb that instructors see so much of on biennial flight reviews.

While the cut-and-try system of airspeed control in the climb does the job, that is not the whole story by a long shot, at least as far as the effort required from the pilot is concerned. The trouble is trim changes: nose heaviness or tail heaviness in the climb. Before takeoff set on the nose-up, nose-down trim scale what you think will give a not too nose-heavy or tail-heavy airplane with the nose held in the proper attitude in the initial stages of the climb. Or you may, from experience, have learned that for the present CG situation, the trim-indicator needle needs to be just so to give maximum rate of climb speed once you get going. Either way, in the early stages of the climb the pilot is most likely going to have to override what the airplane may be wanting to do as far as nose position goes. Raising the gear in some airplanes makes the nose pitch down a bit, and in some, pitch up, which, of course, is going to change the airspeed unless the pilot in command supplies the necessary forward or back pressure on the control wheel to prevent an attitude change. Or the pilot may do that and at the same time start retrimming the airplane. Or simply hold whatever control pressure it takes to maintain attitude knowing that when takeoff flaps are raised, there's going to be another pitch twitch and he'd rather handle the average of them at one time.

Obviously a pilot can be mighty busy and sometimes seriously distracted in supplying the control pressures necessary to prevent the changes in attitude that can be caused by power and speed and gear and flaps, until he can retrim. Some pilots are constantly on the trim tab in the first several hundred feet of the climb, which may be all right if they maintain attitude control rather than let a prior trim setting change it. My preference for a final-climb trim setting is just a bit nose heavy, thus requiring a light back-pressure on the wheel to hold the attitude that gives maximum rate of climb speed. That way any error in a moment of inattention will be on the too fast rather than too slow side.

TRIM FOR HEADING EASE

Meanwhile, or better, simultaneously, heading control in the climb. Here again takeoff trim setting varies the demands on our attention. After lift-off we either try to keep the airplane on the runway heading or go to a crab heading if there's a crosswind. The propeller's corkscrew slipstream effect on the vertical fin in a single as well as the downgoing propeller blade's pulling harder than the up-going one on the opposite side make the airplane want to yaw and roll to the left. Maintaining a proper rudder pressure, with the wings held level, can take care of this phase of heading holding, but the amount of rudder pressure necessary is very much affected by the position of the rudder-trim tab. Some pilots prefer to leave the rudder trim set for cruise conditions in takeoffs and even approach rather than have to do a lot of rudder trimming. Others prefer to run in a bit of right-rudder trim before the start of the takeoff run, and retrim for cruise climb and cruise. But the main thing, whatever the procedure, is to monitor and do whatever it takes to hold the heading.

Most of our general-aviation aircraft have rudder trim only and use ground-adjustable trim tabs on the ailerons for lateral trim, with the tabs bent for cruise conditions. Such settings may make the airplane wing-heavy in a climb, or it may be wing heavy from fuel imbalance in the wing tanks or placement of the load laterally in the cabin, but in any case the pilot's heading holding will be affected just as it is by rudder-trim tab setting. Some pilots prefer simply to leave the rudder trim set for cruise and hold the right wing down in the climb as necessary to hold the desired heading. This seems to have minimal effect on rate of climb, within limits, and takes care of both the directional and lateral trim variations.

Maybe at this point the hat should be passed for the poor pilot trying to prove that nothing should ever happen in a climbout if only heading and airspeed are controlled precisely —because as well as sympathy he can, on occasion, need some coins for running his sweat shirt through the Laundromat. Turbulence alone can press him hard.

I have an ATC 510 instrument flight simulator, which I let a number of great-nephews and -nieces fly. They are from eight to twelve years old, and do surprisingly well until I add a bit of nose-up or nose-down trim and turn on the first stage of turbulence. At this point they stop flying the simulator and it starts flying them; heading, airspeed, and altitude control go to pot. And you know why. It takes not only a firm hand on the reins but at times also a nudge with the spurs to keep an airplane flying in a manner other than in response to its trim and power settings.

Engine failure or malfunction in the climbout for reasons other than fuel-system mismanagement? It's a rarity, but, as you know, it often causes pilots to stop thinking. When it happens, we all have the same instinct: namely, to hold our climb attitude or, in a manner of wishful thinking, even to increase it. That, of course, can be lethal. We need to condition ourselves mentally to put the nose down instantly a bit below level attitude so we can look over the area ahead and 45° to either side to find the best place to put it down. Nothing can be worse than holding the nose up and spinning in at this point. Getting on the ground under control is what makes forced landings survivable. And, more often than you can believe, uneventful.

In twin-engine flying we also tend to hang on to the normal climb attitude after losing an engine. We can quickly lose directional control. In a twin on one engine we need to retain a vivid picture of how little above the horizon the nose rides in a climb on one engine, and instantly get it down to at least that attitude, or even better to a level-flight attitude. That done, we've got a lot more time in a twin than in a single to look over the situation ahead, and even if the twin won't climb, the operative engine gives us a most helpful gliding radius at single-engine maximum rate of climb speed.

The turn-back after engine failure or malfunction? That can be discussed along with the four 90° turns we make in our circuits in the pattern, the four potential trouble areas in which we generate most of our serious takeoff and landing statistics. The turn-back after engine failure and these pattern turns are no different.

SAY AGAIN: DON'T TURN BACK

There is no quicker way to stall an airplane than in a steep and tight turn in the usual low-level turbulence. With gust effects it can happen even more quickly. In the engine-failure cases on takeoff the pilot is thinking about a 180° turn back to the airport, but if he overdoes it, he won't get past the 90° mark before losing control, which is just the way it happens in turns in the pattern. In the attempted turn-back, there's an additional factor: the upper wing may get into the bottom of a sharp edge-overrunning gust, resulting in an overbanked situation in which the pilot needs even more desperately to first move the stick forward a bit so as to increase greatly the airplane's roll-rate capability.

The turn. Don't ever forget the turn at low altitude. In our training at altitude we get the impression that an airplane is really difficult to stall in a power-on turn. No one tells us that almost daily for 50 years now, someone has been involved in a loss of control in turning flight at low altitude, resulting in a spin-in or dive into the ground in an incipient spin or spiral. In the takeoff and pattern our turns are started at a relatively low airspeed with stalling speed in easy reach of a little extra back-pressure on the control wheel. And in following the convention of level turns only at low altitude, we sometimes climb as a result of overtightening the turn, which lights an even shorter fuse. We might be better off were we trained to make slightly descending turns in the pattern.

THE PATTERN REVIEWED

Downwind leg? In circuit flying we often create for ourselves the same problem we have in pattern entries: too much speed. Headed downwind after the crosswind leg we often go to cruise power and cruise speed and either wind up with a base leg in the next county or else hurry into a steep turn to keep the base closer in. Biennial administrators say we also don't do too good a job of holding either heading or altitude and, in fact, climb above pattern altitude on downwind.

Base leg is a slack period when surely nothing should ever

go wrong. Unless, perhaps, with still too much speed, we overshoot the runway center line and rack it up into a level, steep, tight turn.

When the FAA decided that with-power approaches were indeed respectable, I think they did so on the basis that the power-off approach requirements led to too many too-long landings, too many too-late go-arounds, too frequent power-off glide-stretching attempts, and because they invited engines to load up in long throttle-closed prop-windmilling finals.

Power approaches, they must have reasoned, should eliminate all of this and give a big gain in safety. Besides, when we trust an engine for hours to get us to a destination, why shouldn't it be even more reliable at partial power in that last mile? I think the reasoning was sound, and the hypothesis still valid, but it doesn't really seem to make the difference it should. Not just among the have-nots but even among the haves. The other day at a controlled field where I do half of my flying, a King Air came in low and fast, touched down level and fast at midfield, and didn't much more than get stopped at the end of the 6000-foot runway. Right behind this one came an Aerostar, which did the same thing. The next airplane to land was another King Air, which came in low, made a tail-low touchdown, and turned off at midfield. And right behind him, so help me, was another Aerostar, which also made the midfield turnoff. There's also a good deal of corporate-jet traffic on this field, and it is not uncommon to see one of them touch down at midfield and get stopped just in time. But theirs is not just a high stall- and approach-speed problem. The tire marks of Piedmont's 20 or so 727 and 737 arrivals here each day are in the first few hundred yards of the runway. We can put ours there, too, if the approach is correctly executed. Which, unfortunately, it often isn't.

FINAL: THROUGH THE THRESHOLD WINDOW

However we get on final, either from a base-leg turn or with a straight-in clearance from approach control, we all have the same problem to solve. The goal is to put the airplane through a window say 50 feet above the end of the runway, flying at the proper approach speed and, in most general-aviation aircraft, with around a 500-FPM rate of descent.

Sometimes we come in too high to get it through the window, even though our speed is right. And sometimes we're too high and too fast to even get it into the airport. At other times we get through the window all right but are much too fast to land in the first third of the runway. At still other times we come through the window too slowly and with a rate of descent so high that it makes the flare critical.

It seems all too obvious to mention, but our so frequent long landings, which wind up in the fence or ditches, or else in a collision with objects other than the ground as a result of too late go-arounds, could all be eliminated were it standard practice to abandon any approach that didn't take us through our window with all our skills showing.

In order to improve our score on going through the window with the right speed and rate of descent, the first prerequisite after turning final is to be sure that, once in landing configuration, we are not carrying any speed in excess of 1.3 times stall speed. Too much speed on final is almost a guarantor of a fouled-up landing. Once we're rid of any excess speed, it is also helpful to get trimmed slightly nose-heavy at approach speed and continue toward the runway at pattern altitude.

At some point we have to make a decision about how steeply down we're looking at the window. The prospective approach path should be shallow enough that we can carry some power all the way down, thereby leaving ourselves a

---

correction margin of either adding or decreasing power. If it is too steep, we may wind up with a power-off approach and its attendant difficulties of precise glide-path control. As we start down, we want the power set to give us 500 FPM down along the approach path with the speed kept rigidly on 1.3 V_{so}. We want staying on the approach path to be a question of more or less power rather than more power or no power. This is what gives us uniform power approaches right through the window.

It is important, too, that once we start down on final we go to full blower in perceptivity. It's easy enough to exercise reasonably precise heading, airspeed, and altitude control in stable air, but on a wild and windy day we need to be highly aware of even the slightest change in attitude of the aircraft. In these approaches the airplane is going to want to yaw, pitch, and roll, sometimes singly and sometimes in all three axes at once. No matter where we're looking—over the nose, or out one side—the yaw, pitch, and roll clues are plentiful and require immediate action. A change in the angle of the bottom of a windowsill to the ground or anything on the ground that is vertical, such as a tree, or corner of a building, means the airspeed is going to increase or decrease unless we do the necessary. The bottom of a wing, or angular movement of a wing strut, or the top of a wing can give the same information. And as well, looking along the wing will give a yaw signal by the wing's fore or aft movement. If a wing's down a bit, that's going to bring on a heading change. So as the ball bounces, whether we're looking for traffic or over the nose straight ahead in a rough sea, we can keep our approach stable in speed and heading only by keeping the airplane from deviating from the attitude that will provide us those ingredients of a proper approach.

This is the visual part on final. We also need to be responsive to any changes in rate of descent. A little lighter in the seat, or a little heavier, or the feeling one gets in an elevator as it starts up or down, indicates a faster or slower rate of vertical travel. Which tells us that if our airplane starts settling, we add power and change nose position to hold the airspeed constant, or if we get into an atmospheric burp, we reduce power and nose down a bit to hold our proper approach speed.

So through the looking glass we go, and if we've done our control job right, we're handed a bonus, which comes from consistency in our approaches: we can always count on the same time sequence from window to touchdown, and that helps. Excess speed in our round-out and hold-off as a result of a too fast approach sentences us to a longer-than-necessary exposure to the vagaries of the atmosphere.

But, alas, coming through the window with all the virtues doesn't eliminate the level-off/touchdown skill required. If we're flying an airplane with a stiff gear, it helps to complete the round-out only a few feet above the runway. That way, if we're going to drop it in, we haven't far to go, and the gear's reaction will be less severe.

DOES IT FLOAT?

If we're flying a low-wing airplane with the wings unusually close to the ground when the airplane is parked, we may have a floating characteristic to handle. In these, leveling off as high as 10 feet above the runway and letting the airplane settle into the ground cushion as the nose is raised seems to cut down noticeably on the distance the aircraft floats before touchdown.

Our most common problem may occur on a windy day when we get rounded out and start thinking in terms of holding our attitude until it starts settling and then from there on raising the nose at the right rate to keep the rate of settling quite low. Ideally, we would give it the rest of the back travel of the control wheel when we get the feeling that the wing is fixing to let go. This pleasant idyll can be abruptly ended by our flying into a gust. Next thing we know, we can be 12 feet above the runway instead of only six. Or even 15 feet.

This is the sort of thing that causes the FAA to insist that we must do our approaches and round-outs and landings with one hand on the throttle. They feel that at the top of the gust effect there's no time to spare reaching for the throttle. The power needs to go up some, the nose down a bit, and having regained whatever speed has been lost in the ballooning, we then start the last phase of our approach all over.

A hand always on the throttle may be logical enough, but I don't think it should be grounds for disqualification if the pilot is able to use other means to handle the situation. Namely, when we start to balloon from a gust effect, it is surprising how quickly an instant, slight lowering of the nose pulls the gust's cork. And if the gust is unusually powerful, putting the nose down far enough to stop the ascent takes care of things. Once out of the gust, you're still close to the ground and will have lost a little airspeed but a little faster flare for touchdown will take care of that.

FLY THE WING

There's also another reason for flying the wing and not the engine in the level-off/touchdown phase, and that is the previously discussed high stick-force gradient of nearly all our four-place-and-up airplanes today. With approach trim unchanged in the flare and touchdown, it takes so much back pressure on the control wheel that most of us need both hands on the wheel to get a good touchdown. The load is so high that, pulling for all you're worth with just one arm, you lose too much feel for the speed and lift situation. I think the hand-on-the-throttle question should be left to the pilot. The rule evolved at a time when all airplanes were taildraggers and dropping one in on the mains only could produce a lively zoom into an incipient stall 30 feet above the runway, from which a power recovery was a necessity. If landed even hard on the mains, our tricycles can bounce, but rather than pitch up they tend to pitch down because of the location of the mains behind the CG. The hand-on-the-throttle rule has long since served its main purpose and is out of date.

Now we've come through our window with the throttle closed, rounded out, stopped our descent, and hesitated a moment waiting for the airplane to start settling. When it does, we remain highly sensitive to any rapid, as well as too slow, rate of descent and meter our back-pressure changes on the control wheel accordingly. In doing this we offset any changes in wind velocity or gust effects as we near the runway.

With the nose finally up at least enough to be certain the nosewheel isn't going to touch first, the moment of truth approaches. We start holding it off—that is, we stop the rate of descent with the wheels maybe only a couple of feet above the runway.

What do we do if the airplane starts drifting sidewise at this point? If it's early in the hold-off phase, we make a crossed-control yaw into the crosswind—enough to stop the drift—and, just before touchdown, cross-control it back to runway heading for a no-drift touchdown. But if the drift starts late in the hold-off, we make a crossed-control, wings-level change in heading in the direction of the drift. This may not stop the drift entirely, but it at least gets the airplane pointed more nearly where it's going and any drift transients are likely to be minor. Just before touchdown on the drift heading, lowering a wing slightly so as to get the upwind wheel on the runway first is a form of dragging one's foot to brake any remaining drift.

It has been my observation that taildragger pilots pay more attention to avoiding a touchdown with drift than do tricycle pilots. They have to because such a touchdown is a powerful springboard for a fast entry into a ground loop. Drift is less bothersome to a tricycle pilot, but only to a point. Pilots often seem not to pay enough attention to keeping the wings as level as possible in the last moments of a landing. The second we touch down, the laws of aerodynamics fade out and the laws of physics become effective. As we have all found, in a too fast turn taxiing downwind in a tricycle we can experience having the airplane start to nose over, rotating about its nose-wheel/downwind wheel axis. This same thing can happen, if I may repeat once more, when a pilot, landing with drift, tries to stop the nose-over with coordinated rudder and aileron movement. What is needed at this juncture is a crossed-control maneuver with the nosewheel, now the more powerful lateral control, turned in the direction of drift.

When to close the throttles in the last phase of an approach in a twin-engine airplane is a different decision from what it is in singles. In some of the earlier twins with split flaps, check pilots always put considerable stress on leaving on a little power until the airplane is on the ground and rolling. They caution that closing the throttles any sooner can result in a drop-in and hard landing.

In any twin, windmilling props cause a considerable disturbance in the airflow over the wing area behind the props, and consequently even with trailing edge flaps it's better to touch down with some power on. But there are exceptions to this. For one, the Seneca II. If even just a bit of power is left on in the hold-off phase, it will fly and fly and fly before touching down. Closing the throttle in the approach when you would in a single has no ill effects.

Now we're touching down softly and experiencing the ultimate joy in flying. There is, of course, the roll-out phase remaining, the ground-loop/swerve charade, but no-drift touchdowns take care of that.

COME FULL CIRCLE

The question, again. Instead of giving us the bulk of our misadventures in flying, why should there ever be any untoward events in takeoffs and landings? There shouldn't be. It's entirely a question of adequate airspeed, heading, and altitude control.

We can roll straight enough on takeoff to at least stay on the runway. We can climb at maximum rate of climb speed. We can keep our pattern turns no steeper than 30°. On downwind we can get our speed down to approach speed plus whatever speed loss will result from putting down the gear and flaps. On final we can get into an approach stabilized in speed and rate of descent and select an approach path that can be controlled with power and that will take us through the runway's window. And we can touch down tail-low without drift, in the first third of the runway, and roll to a stop.

And wish no flight ever had to end.

INDEX

Climbing (Cont.)
of attack; Basic concepts; Take-
off; Winds
Collisions, 167–68; with ob-
structions, 171
Comanches, 44, 118–19, 227.
See also Twin Comanches
Conklin, Ed, 98–99
Conquest (Cessna), 257
Consistency, in design, 81
Constellations (Connies), 54,
135
Control wheel, 93–94 (*See also*
Stick); basic concepts, 5
Courier. *See* Helio Courier
Crabbing, 19, 219–20ff.
Crosswind landing gear, 222–24
Crosswinds, 19, 216–18ff., 224–
27, 277, 290–91
Cruise climb, 156
Cubs, Piper, xiv, 31, 207
Culver Cadet, 127, 180–81,
194–95
Culver V, 270
Curtiss, Glenn, 57–58
Curved takeoff run and landing
roll, 225–27
Cutlass, 34, 35, 115

D

Dayton, Ohio, 102
DC-3's, 188, 209
Density altitude, 157, 249–50ff.
Denver, Colo., 54
Design. *See* Aircraft flight char-
acteristics
Detroit News, 203

Direct linkage, 80–81, 82
Dives, 37–38, 175–76
Downdrafts, 60, 174. *See also*
Undershooting
Downsprings, 104–11. *See also*
specific aircraft
Downwind leg, 164–65, 166,
184
Drift, 216ff., 227–28, 290–91.
See also Crosswinds; Ground
loop/swerve
Dual systems, 259–60
Duchess, 256, 257–58, 265
Duke, 256

E

Eighth Air Force, 133–35
Elevators, 270 (*See also* Airspeed; Altitude; Angle of attack; Downsprings; Trim;
specific aircraft); basic concepts, 7–8, 11, 16; breakaway
friction, 94–96
Engine failure/malfunction,
158–60, 182–83, 212, 282–83
(*See also* Stalls; Twins; specific
maneuvers); fear of, 130–31
English Eighth Air Force,
133–35
Ercoupe, 18–21, 50, 135, 261,
269

F

FAA, 43, 143, 174
Fairchild, 24, 95

Humidity, 153–54
Hunt, Woody, 42
Hydraulic steering, 83–84

I

Ice, xv, 2, 159
IFR. *See* Instruments
Initial climb, 156
Instruction. *See* Training and instruction
Instrument panels, height of, 78–79
Instruments (IFR), xv, 1, 2, 99ff., 185, 187–90. *See also* Twins
Intuition, 128–29

J

Jato bottles, 260
Javelin autopilot, 102
Jets, 285. *See also* Airlines and airliners; Twins; specific aircraft
J-3, 76, 207

K

Kelley, Bill, 74–75
Kennedy Airport, 52
King Air, 256–57, 285
Kitty Hawk, 224
Koppen, Otto, 13, 205
Kurt, Franklin, 224–27, 258

L

Lance, 120, 152
Landing, 161–220. *See also* specific aircraft, maneuvers
Landing gear (*See also* Gear-up landings; Taildraggers; Tricycles): stability, 84–85; stiff, 119–20
Langewiesche, Helen, 7–8
Langewiesche, Wolfgang, 7, 209–10; foreword by, xi–xvii
Langleys, 251
Lear, Bill, and Lear, Inc., 98, 101, 102, 185
Learjets, 118, 257
Lester, Ken, 180–81
Level-off touchdown, 193–212, 288–92
Liberators, 133–35
L-19, 204–5
Loading. *See* Weight and loading
Lopresti, Roy, 110
Luscombe, 150

M

Martin, Derwood, 115, 116
Mental attributes, 129–36
Midair collisions, 167–68
Mitchell, Don, 102
Monocoupe. *See* Velie Monocoupe
Mooney, Al, 77
Mooney Aircraft, 14, 77, 109–18, 269–70
Mushing, 21, 160, 175

N

Nagel, Frank, 135
Navajos, 34, 256
Navy, 153, 250–51, 270
New York, Kennedy Airport in, 52
Night, 1, 2, 156, 157, 186–87
Numbers, runway, 184–86

O

Obstructions, 171
Oil, 241, 246
1,000 Day Battle, The, 133–35
Overshooting, 177–80ff. *See also* Twins

P

Panic, controlling, 135–36
Pattern. *See* Traffic pattern and circling
Pedals, 79–80
Philadelphia airport, 53–54
Physical attributes, 123–29
Piedmont, 287
Pilots, xvi, 1–2 (*See also* Training and instruction); characteristics, 123–44; and twins, 245–47
Piper, Howard, xiii–xiv, 195, 202, 248
Piper, Thomas, 207
Piper company, 76, 87, 100, 194, 256; Aztec, 227, 246,

249, 253–54; Chieftain, 256; Comanche, 44, 118–19, 227 (*See also* Twin Comanches); Cub, xiv, 31, 207; and direct linkage, 80; Navajo, 34, 256; Seneca II, 87, 118, 256ff., 291–92; Tri-Pacer, 200; Warrior, 34, 35–36, 91
Pitch, 111–12, 287 (*See also* Autopilots); basic concepts, 5, 6, 16
Pitts, 14
Pontiac, Mich., 185–86
Power (*See also* Airspeed; Engine failure/malfunction; Throttle): when to chop on takeoff, 154–56
Power-off landing, 116–17
Propellers, windmilling, 237, 238, 291
Propeller-unfeathering accumulators, 271–72
Push-pull, 260–61

R

Radio broadcasts, 168
Relative wind, 48–49. *See also* Winds; specific effects
Robertson conversions, 206–7
Roché, Jean, 81
Rockwell, 223, 257
Roll, 287 (*See also* Ailerons; Landing; Turns); basic concepts, 5, 6, 16
Rome, 55
Rough/soft field operations, 151–53, 154, 157

Rudders, 87–93 (*See also* Steering; Trim; Yaw); basic concepts, 5, 12–21; and direct linkage, 80

Runways (*See also* Landings; Level-off touchdown; Rough-/soft field operations; Steering): numbers, 184–86

S

Seminoles, 256, 257, 265
Seneca II, 87, 118, 256ff., 291–92
Service ceiling, 249
707's, 54, 55
727's, 174, 285
737's, 285
747's, 19, 54, 208
Shrikes, 83, 234–38ff., 248
Sideslips, 179, 217–18, 219
Sight, 124–25. *See also* Visibility
Sikorsky, Igor, 128
Simplifly lever, 270
Sinking (sink rate), 138, 173–74ff., 179. *See also* Overshooting; Undershooting
Sixth sense, 128–29
Skids, 17
Skyfarer, 269
Skylane, 98, 180, 200, 227
Skymaster, 261
Smell, 127
Snow. *See* Rough/soft field operations
Soft field. *See* Rough/soft field operations
Sound (hearing), 126

Sperry, 101
Spins, 31, 36ff., 45, 63, 136–38, 143, 167, 173–74. *See also* Stalls
Springs (*See also* Downsprings): rudder/aileron, 87–93; steering, 81–82
Stalls (stall/spin syndrome), 18–21, 25–26, 31ff., 45, 63, 66–69, 142, 143–44, 160–61, 163–65 (*See also* Angle of attack; Engine failure/malfunction; Level-off touchdown; Spins; Turbulence; Turns; Twins; specific aircraft, equipment); G-STALL indicator, 272
Steering, 79–80, 81–85ff., 153. *See also* Center line; Rough-/soft field operations
Stick (*See also* Control wheel; specific uses): basic concepts, 7ff.; pumping, 95–96
Stick-force gradient. *See* Downsprings
STOL approach, 204–7
Storms. *See* Thunderstorms; Turbulence
Story of the Winged S, 128
Stout, William B., 81
Strohmeier, William D., 118, 119
Strohmeier system, 118–19
S-turning, 180
Super King Air, 257
Swerves, 17. *See also* Ground loop/swerve
Swift (Globe), 152

T

Tactair, Inc., 42–43, 269
Taildraggers, 17, 84–85, 150, 151, 199, 210, 212, 276, 277. *See also* Landing; Takeoff; specific aircraft
Tailwind, 56–57. *See also* Winds
Takeoff, 145–61. *See also* specific equipment, maneuvers
Taste, 127
Taxiing. *See* Ground loop/-swerve; Landing; Steering; Takeoff
Teledyne Instantaneous Vertical Speed Indicator, 189–90, 270
Temperature. *See* Density altitude
Thermal effects, 61–63
Thinking, 147–48
Thompson, Bill, 261
Throttle (*See also* Airspeed; Altitude): basic concepts, 7–8ff., 10–11; stops, 194–95
Thunderstorms, 54, 63, 64
Tiger, 82
Time, sense of, 128–29
Touch (feel), 125–26
Traffic pattern and circling, 163–68, 284–88
Training and instruction, 141–44 (*See also* specific aircraft); for twins, 241–44
Tricycles, 17, 85, 149–50, 151, 176–77, 199, 276ff. (*See also* Landing; Takeoff; specific air-craft); bent nose gear on, 200–201ff.
Trim, 39–43, 96–103, 111ff., 175–77, 269ff., 276ff., 281–83 (*See also* Aircraft flight characteristics); three-axis, 96–101ff.
Tri-Pacers, 200
T-33's, 44
Turbo Commander, 257
Turbos, 253–59ff.
Turbulence, 63–69, 157, 259, 279ff. *See also* Gusts
Turn-backs, 247, 283–84
Turns, 12, 15–16, 25–36, 28–36 (*See also* Turn-backs; specific equipment); weights, measures, and, 32–36
TWA, 209
Twin Beech, 260, 271
Twin Comanches, 118, 119, 140, 227, 242–43, 252, 271
Twins, xv, xvi, 27, 139, 155, 156, 158, 172–73ff., 195, 231–67, 283 (*See also* specific aircraft, maneuvers); contrarotating props, 15; rudder/aileron interconnect springs, 91–93; and three-axis trim, 99; windmilling props, 237, 238, 291–92

U

Undershooting, 172ff. *See also* Twins
Updrafts, 61–63, 279. *See also* Overshooting